THE POLITICS OF DEVELOPMENT

Transportation Policy in Nepal

Monograph Series No. 22

Aran Schloss

Center for South and Southeast Asia Studies
University of California
Berkeley, California

UNIVERSITY PRESS OF AMERICA

LANHAM • NEW YORK • LONDON

University Press of America,™ Inc.

4720 Boston Way
Lanham, MD 20706

3 Henrietta Street
London WC2E 8LU England

Printed in the United States of America

Library of Congress Cataloging in Publication Data

Schloss, Aran.
The politics of development.

Bibliography: p.
Includes index.
1. Transportation and state—Nepal. 2. Highway planning—Nepal. 3. Roads—Nepal. I. Title.
HE269.9.S34 1983 388.1'09549'6 83-6807
ISBN 0-8191-3250-0
ISBN 0-8191-3251-9 (pbk.)

Co-published by arrangement with the
Center for South and Southeast Asia Studies
University of California, Berkeley

The Center for South and Southeast Asia Studies of the University of California at Berkeley is the coordinating body for research, teaching programs, outreach and special projects relating to South and Southeast Asia in the University of California system. The Center publishes a Monograph Series, an Occasional Papers Series, and a Language Teaching Materials Series. In addition, it sponsors the publication of books on South and Southeast Asia by the University of California Press.

Publication of this monograph was aided by a grant from the Harry S. Truman Research Institute, the Hebrew University of Jerusalem.

Manuscripts for consideration should be sent to: Editorial Committee, Center for South and Southeast Asia Studies, 260 Stephens Hall, University of California, Berkeley, CA 94720.

To my mother and to Boby

FOREWORD

The debate in Asia, and particularly South Asia, over economic development models has become more intense in the past decade since it has become apparent that the development strategies introduced in the 1950s and 1960s have not had the results predicted by their advocates. Despite the rhetoric used, the issues in dispute are no longer primarily ideological as regimes impelled by basically different ideologies have had similar results in many instances from quite different development programs. An economic program may work reasonably well in one case and be an abject failure in another, which raises the question as to whether it was the "strategy" or some specific cultural-social-political factors that determined the result. The ongoing debate still focuses on whether "trickle down" growth stategies—which have had some spectacular successes in Asia as well as some equally spectacular failures—or "social justice" strategies directed at the redistribution of wealth are best suited for rapid economic development. While these two strategies may not be mutually exclusive economically, they have proven to be difficult to combine politically.

Nepal provides a particularly useful case study on this issue as Kathmandu has rigorously applied a "trickle down" approach in its most basic form since the 1950s. A heavy proportion of government expenditures on development has been directed at basic infrastructure projects—communications and hydropower in particular—on the assumption that these were a necessary precursor to economic growth in agriculture and industry in a difficult mountainous environment. The politically necessary concessions to "redistributive justice" program have, by necessity, been nominal and peripheral.

Aran Schloss, in this thorough and well-conceived study of both the economic and political consequences on Nepal's development strategy, has made a major contribution to our understanding of its weaknesses and strengths. The results to date have had a minimal impact on the livelihood of most Nepalese; but it is also evident that _no_ development strategy would have done any better without these basic infrastructure programs. Kathmandu made the right choice in the 1950s; it is only now that alternative options on growth models are feasible. Moreover, as the policies adopted involved a heavy dependence on foreign assistance, Schloss has analyzed in some depth both the impact of such dependence on decision-making in Nepal and compared the relative effectiveness of the different approaches to aid programs adopted by the different foreign donors. This study, thus, is very relevant to those Asian societies that, by necessity, require external support for their development programs as well as those governments and international organizations that provide assistance.

Leo E. Rose
Political Science Department
University of California
Berkeley, California

PREFACE

Nepal began its journey toward modernization in the early 1950s. It continues to follow this path and tries to arrive at solutions that will be congruent with both its heritage and the promise of modernization. The monarchical regime of this Hindu Kingdom guides the country's efforts to develop, seeking to introduce gradual changes that will enable economic growth while sustaining the present form of government. Nepal which is among the least developed nations in the world achieved some discernible progress in several sectors of its economy but is yet to break through the "vicious circle" of poverty. In its development efforts the country embarked on a variety of programs among which the establishment of a national transport network occupied a central place. In fact, among all nations Nepal stands at the top of the list for the sizeable investment it made in this particular sector. In this effort the country succeeded in mobilizing considerable assistance from several aid-donor countries which enabled it to have its primary road network established by 1976. The emphasis put on this sector and the enormous investment funneled into it constituted an almost single sector strategy of development. When combined with the multitude of participants and the variety of political and economic interests involved in it, the development effort in the transport sector represents most clearly the dynamics of the development process in Nepal.

This study examines the planning and the implementation of development programs in the transport sector of Nepal. The study analyses these processes in terms of the interactions that evolved between the factors active on the Nepali scene. The analysis centers on the dynamics of Nepal's indigenous political-administrative system and on the political and economic interests pursued by the foreign participants as well as on the administrative practices they employed in this process. The study therefore deals with the dynamics of development administration and development policy-making and the results they yielded in Nepal.

Organization

This study is organized into two major parts. The discussion opens with an introductory chapter which deals with the interrelationships between politics, bureaucracy and development. This review discusses the implications of these interrelationships on development programs in the transport sector in general and in Nepal in particular. Chapter 2 reviews the history of the transport sector in Nepal during the years 1951-1975. This presentation consists of a detailed review of the priorities given to the transport sector in four consecutive Development Plans. Secondly, it includes a discussion of the main components of Nepal's Master Plan for roads, followed by a general description of the roads actually constructed in the country. It concludes with an examination of the allocations and actual expenditure in the transport sector and the progress attained in road construction operations during the 1951-1975 period. Chapter 3 deals with the dynamics of the political and administrative processes in Nepal. It examines the evolution of the country's political-administrative system with particular reference to the effects it had on the function of planning for development. The discussion is divided into three sections, each corresponds to a distinct stage in Nepal's more recent history—a) the decade of political experimentation, 1951-1960. b) the royal coup and the establishment of Panchayat democracy, 1960-1967 and, c) the consolidation of the crown's power, 1967 to the present.

The second part, chapters 4 through 7, opens with a proposed framework of analysis to deal with the development process in Nepal. This framework identifies the factors involved in the planning and the implementation of road projects and presents a set of hypotheses on the interactions between them. Chapter 4 discusses the dynamics of the planning process in terms of these factors and in the context of the interplay between the internal arena and the external participants. The role, the interests and the actions of these participants in this particular field in Nepal are examined together with the means and ways utilized by Nepal in obtaining aid and in pursuing its development objectives. The outcomes of this interplay are discussed in terms of which roads were built by whom and where. Chapter 5 further elaborates on this process and relates between the planning process and the preparations made toward the execution of the project. This chapter deals first with the manner in which Nepali agencies performed this task in the projects they undertook to construct. Second, it examines the manner in which five major foreign participants—China, India, USA, USSR and UK—prepared their respective projects for execution and the patterns of interaction they maintained with their Nepali counterparts. Chapters 6 and 7 deal with the process of implementation. Chapter 6 examines the dynamics of Nepal's indigenous political-administrative system as they came to bear upon the execution of road projects by Nepali agencies. This includes the analysis of organizational communication, practices of coordination, the application of administrative control mechanisms and the management of personnel. The analysis discusses the reasons behind this mode of operation and the impact it had on the execution of projects. Chapter 7 which concludes the discussion deals with the interactions that evolved between the Nepali agencies and the foreign participants. This chapter identifies the challenge of foreign aid activities and the requirements that it presents to aid-donor countries: organizational flexibility, balance with the project's environment and balanced relationships with the partner-client—i.e. the recipient. The discussion then proceeds to compare the patterns adopted by the donors to meet this challenge, followed by the analysis of the donors' record in the execution of their projects and the responses of the Nepali participants to these modes of operation. The discussion concludes with a brief summary on the process of implementation in Nepal.

Data

The data on which this study is based was gathered in Nepal and in the USA. The author carried out his field research in Nepal during the period November 1975 - July 1976, in the course of which some 160 interviews were conducted. Among those interviewed were Nepali officers whose positions ranged from Cabinet Members to middle level administrators. Nepali political personalities, intellectuals and some 30 foreign experts (from Western aid-donor countries and multilateral agencies) were also interviewed as were staff members of several diplomatic missions in Kathmandu.

Numerous internal reports and studies by international organizations and aid missions were obtained in Nepal and abroad, and they include several confidential documents, the source of which will have to remain undisclosed. Reference to the literature and documents used is given at the end of each chapter.

ACKNOWLEDGMENTS

It is a pleasure to thank friends, colleagues and acquaintances for their help in this study. I am most deeply grateful to Prof. Leo Rose for his counsel, his exceptionally helpful comments, his kindness, help and encouragement in all stages of the study both in Nepal and the USA. Many thanks are also due Prof. Martin Landau, Prof. Lovell Jarvis and Prof. John Scholtz for their valuable comments.

The help and kindness I received from many people in Nepal made the field research a most rewarding and pleasant period. To all the Nepali officials and private citizens who shared with me their knowledge of the system I am greatly indebted. Their request to remain unidentified is respectfully honored. I would also like to thank the foreign experts serving in Nepal, mostly with USAID and UNDP, for their kind assistance, and in particular to Bernie Hausner, UNDP's Assistant Resident Representative.

My special thanks go to the "family"—Lidya, Yair, Atida, Shaul, Ronnie and Giora who made it so much easier and pleasant.

My field research in Nepal was supported by fellowships from the Institute of International Studies, UC, Berkeley; The Institute of Urban Studies, UC, Berkeley; and the Hebrew University of Jerusalem, Israel. I am most thankful to each of them for their kind assistance. I would also like to thank Prof. Z. H. Schifrin for his counsel and the H. S. Truman Institute, Hebrew University for its support.

Thanks are due to Marcia N. Doron for her assistance in editing the manuscript. To N.M., S.L. and A.H. I am indebted for their encouragement and patience.

Finally, my family has been a source of constant help and encouragement which made it possible in every corner of the globe all the time.

TABLE OF CONTENTS

Chapter 1

POLITICAL AVENUES AND ADMINISTRATIVE ROUTES IN THE FIELD OF TRANSPORTATION FOR DEVELOPMENT

Development, Politics and Bureaucracy

The easy way for a nation to attain sound, well-balanced economic growth within a short period of time is yet to be found. The future prospects for such an accomplishment seem more and more remote as the complexities of the development process are being realized. The increasingly larger volume of knowledge on the factors involved in the development process enables us to identify some inadequacies of past solutions. It also points out the insufficiency of existing knowledge in overcoming the numerous problems of development. Hence, we must further pursue our quest for a solution to the issue of development.

Students and practitioners of economic growth and development agree that modernization is a complex process which cannot be reduced to a single factor or to a single dimension.[1] Indeed, it became apparent that huge infusions of foreign capital into poor countries neither produced the expected "sustained growth" nor enhanced necessarily the immediate capabilties of most of those countries to reach the "take off" point. It was increasingly realized that being poor meant more than lack of sufficient monetary resources. The list of the characteristic feature of a poor country expanded to include the technical and non-technical aspects of the matter. Poor education, limited welfare and health services, subsistence economy based on primitive agriculture, low rate of investment and high rate of population growth, lack of means of communication and insufficient transport facilities were pointed out as the prevailing physical circumstances in the poor countries. Lack of capable manpower, the insufficiency of data and limited administrative capacities were identified both as an outcome of and a cause for "sustained poverty." Cultural attitudes toward change, the introduction of new and mostly foreign concepts and organizations into society and, the viability of the so-called traditional institutions were depicted as crucial factors that enhanced disagreements over the course and pattern of development within each of those countries and contributed to the political instability that plagued most of the LDCs. External political and economic interests and the ideological differences among aid-donor countries were identified as elements that had direct bearings on the development process of each LDC, both in terms of their contribution to it and in terms of the impediments they created in it. The issue of development and the problems of the poor countries were defined in increasingly complex terms, and the larger the number of factors identified as partaking in the process or impinging upon it, the more difficult it became to find adequate solutions to those problems.

Students and practitioners of economic growth and political development also agree that the results attained thus far on the development scene were not particularly encouraging. Though they differ in the assignment of their priorities, most seem to consider the "bureaucratic machinery" among the major bottlenecks to development. The term connotes a decision-making structure and process that reduce the prospects for effective planning and the efficient implementation of development programs. It includes the deficient setting out of priorities, excessive red tape, flourishing corruption, ascriptive criteria in every organization and misusage of administrative authority. The term also implies bureaucratic incompetence, ineffective control mechanisms combined with a great emphasis on control-oriented procedures, deficient communication among administrative units and a high degree of uncertainty within the bureaucracy. Though the causes for such phenomena are many and involve the cultural, administrative, technical and economic dimensions of each LDC, the very consistency of the problems is largely attributed to the pattern of interaction that evolved between politics and bureaucracy, between the political process and the administrative process in the LDCs.

The emergence of the administrative machinery as an active participant in the political process and as an arena where a large part of the competition over distribution of power took place was, in most cases, the result of the problems that confronted the LDCs. It involved scarcity of resources, shortage of capable manpower, colonial heritage, dependence on foreign aid, lack of communication systems and transport networks, insufficient data, traditional group affiliations, and limited administrative capacities. The "modern" bureaucracy, whether inherited from the colonial country or imported from another country by means of foreign assistance, was adopted as the tool to solve those problems and to carry out the tasks of development. However, this adoption often proved superficial. The principles behind the functions of the bureaucracy were rarely carried over; adjustments that were required in this essentially foreign institution often proved insufficient to make it more congruent with the prevailing socio-political and cultural properties of the "borrowing" country. Instead, these adjustments and the manner in which most groups in the society came to perceive the bureaucracy made the administrative machinery a "natural" candidate for participation in the political process. This institution was given control and authority over resources and consisted of some of the more educated segments of the population. It could hardly escape the "fate" of becoming a target of pressures from groups in the society as well as a self-conscious elite expected to approach and deal with any given issue from a broad national development perspective. The administrative machinery, either explicitly or implicitly was assigned the role of a major agent for change, with vaguely defined limits. Mechanisms that could presumably monitor the actions and inactions of the bureaucracy (e.g. viable national political organizations, articulation of interests, communication facilities, regulative and distributive capacities of the political system and, a basic consensus on means and ways for development) were either non-existent, too weak, or influenced by the bureaucracy. Many LDCs witnessed therefore the emergence of a bureaucratic machinery that became increasingly involved in the performance of functions usually associated with political organizations per se. Furthermore, once the bureaucracy gained these attributes, it became more difficult for all concerned to dismantle its powers. This is particularly true under circumstances of political instability and slow progress in the development effort (for which the bureaucracy was partly responsible). It was suggested that "Poverty is not merely a condition to be overcome, but the major cause for its on continuance."[2] Similarly, the politicized bureaucracy evident in most LDCs was not merely a constraint to be dealt with, but an important factor in its own preservation. The politicized bureaucracy largely contributed to the continuous state of limited administrative capacities, which in turn affected the country's performance on the development arena and subsequently enhanced the conditions that sustained the bureaucracy's role in politics.

This study deals with the interplay that evolved between politics and bureaucracy in one developing country in Asia. It analyses the ways the country's leadership and politically active groups used the administrative machinery to further their political interests. It examines the political roles and functions the bureaucracy acquired for itself as an active participant in the political process. The study focuses on the impact that this interplay between politics and bureaucracy had on Nepal's development policy-making and on the planning and implementation of the country's development programs.

Transportation, Development and Politics

It is axiomatic that adequate and effective transportation facilities are a necessary condition for development. Such facilities affect the pattern of economic activities, the nature of production practices, the growth of industry and the emergence of new enterprises. Adequate transportation networks enlarge the market and introduce changes into the marketing system as well as new patterns of consumption. They enhance the tapping of new resources and stimulate economic specialization. Adequate transportation facilities affect the infusion and spread of development ideas and provide the means for more social interaction. They contribute to changes in forms of political and economic demands, and largely determine the effectiveness of governmental presence in remote areas and services rendered, including law and order. They also bear upon strategical considerations, internally as well as externally, in terms of accessibility, ground movements and military operations. Expanded transportation facilities, although not a sufficient condition for either economic growth or fuller political participation, are undoubtedly a major factor in enhancing such processes.[3]

A review of the literature on transportation and development reveals that most studies were conducted by students of economic development and transportation engineering. Those studies concentrated on the economic and technical-technological aspects of the issue.[4] Furthermore, of the few studies that dealt with the political and administrative dimensions of transportation, the large majority concentrated on subjects related to international politics and military strategies (historical cases like the Sea-Road in the Mediterranean, Hannibal, Alexander the Great, the British and Spanish Empires as well as more recent cases like the Vietnam War, port facilities to NATO forces and the Russian Navy in the Mediterranean Sea, to mention on a few examples). A somewhat greater interest in the issue of internal politics and transportation could be found in those studies that deal with the question of transport facilities in rich countries and particularly so with respect to urban transportation. The issue of political and administrative factors in transport projects undertaken in the LDC was apparently widely neglected by the literature. This applies to both the impact of these factors on the planning and implementation of transport projects and to the question of the impact such projects had on the political arena (i.e. political consequences emanating from inauguration or completion of projects). This "missing part" is rather surprising since the literature on development and modernization stressed, time and again, the important role of political considerations and administrative factors in determining both the content and the outcome of various development projects.

It seems therefore reasonable to assume that transport projects are also affected by these factors and have their own impact on the latter. If, however, this proves not to be the case, then transport projects constitute a significant exception to the rule. In both cases the issue deserves further examination, particularly since transport projects usually require large investments and as such attract the interest of politically active participants. Secondly, since most transport projects in the LDCs were undertaken by aid-donor countries, and, since foreign aid programs are rarely given for altruistic reasons alone, it can be assumed that donors and recipients alike applied political criteria in either the planning or implementation of these projects.

This study discusses the political and administrative forces that impinge upon the planning and implementation of transport projects. The study analyses the experience of a country (e.g. Nepal) which has invested in the past 25 years more than any other country in the world in the development of its national transport facilities, particularly in road networks. Moreover, in Nepal all major road projects were undertaken by aid-donor countries primarily India, China, the USA, USSR and the UK.

Politics, Transportation and Development in Nepal

Though unique in many respects, Nepal is a typical developing country. GNP per capita is approximately $100. GDP rate of growth per annum is approximately 2.2% and population rate of growth is estimated at 2.2%. Total population is estimated at about 13 million, and crude birth rate and crude death rate (per 1000) estimated at 42 and 20 respectively. Infant mortality (per 1000 live births) is estimated between 200 and 300. Adult literacy rate is approximately 11% and primary school enrollment is estimated at about 35%. The percentage of the population that has access to piped water is estimated at about 7%. Agriculture accounts for about two thirds of the GDP and for 80% of export earnings and provides employment to more than 90% of the population. The urban population is about 3% of the total population; it is mostly in the urban centers that the population has access to electricity (though in some parts of the country particularly in the area around Kathmandu, the surrounding villages also have access to this commodity). In Nepal's First Five Year Plan (1956-1960) the percentage of external resources (foreign aid and loans) in the country's development budget was about 90% whereas in the Fourth Plan (1970-1975) it amounted to 45%. The country's area is 141,000 square km and in 1950 it had a total of 376 km of roads (excluding foot rails) of which 88 km were all-weather roads and 288 km (76%) fair weather road. In 1976 Nepal had a total of 3173 km of road, about half of which were all-weather roads. Industry in Nepal is still at an incipient stage and the manufacturing sector's contribution to GDP is estimated at about 3-4% (in 1966 it was estimated at little more than 1%). Indeed, Nepal has all the characteristic features of an LDC only more emphatically.[5]

Nepal's modern history dates from the latter half of the eighteenth century when the tiny principality of Gorkha, under the leadership of King Prithvi Narayan Shah expanded its domain and brought more than seventy-five hill principalities under its rule. The area King Prithvi Narayan Shah conquered and unified constitute much of present day Nepal.[6] The Shah dynasty continued to rule the country, though it became increasingly weakened by intrigues and counter-intrigues in the palace court. In 1846, following the "Kot Massacre," Jang Bahadur Rana was appointed (by the Junior Queen, Lakashmi Devi) the Mukhtiyar with the title of Prime Minister and Commander in Chief. In a matter of ten years since his appointment Jang Bahadur established himself as the undisputed master of the country and secured a political arrangement under which he and his family were invested with hereditary rights in perpetuity to absolute authority in Nepal (August 6, 1856). This was the beginning of a century of Rana rule over the country with the Shah Kings being kept in the position of political nonentities. The Rana regime was an undisguised military despotism of the ruling faction within the Rana family over the King and the people of the country. The government functioned as an instrument to carry out the personal wishes of the ruling Rana Prime Minister and its main domestic preoccupation was to enhance the personal wealth of the Rana family. No distinction was made between the personal treasure of the Rana ruler and the government treasury. Moreover, the Rana regime pursued a policy of isolation from the outside world while in its domestic policies it adhered to the principle of "status quo" which actually meant lack of development in most sectors of the economy (with few exceptions in the agricultural sector). The system continued fundamentally unchanged, in its basic features and in its reliance on force and coercion until its overthrow in 1951.

In the 1940s, following decades of conflicts and disunity in the Rana family, the ground on which this regime stood began to shake. In 1946-8 groups of Nepali political exiles (in India) established political parties, the largest of which was the Nepali National Congress. The Anti-Rana movement grew larger and louder and demands for changes in the political system were heard from different quarters in and outside Nepal. The concessions that the last two Rana rulers (Padma Shumshere and Mohan Shumshere) were willing to make proved insufficient and tardy to satisfy the growing opposition. Following a series of events (which included the flight of the royal family to the Indian Embassy in Kathmandu and from there to India) and unfruitful negotiations between the Nepali Congress party and the Rana rulers, King Tribhuvan and the political parties joined forces and ousted the Ranas, with Indian backing and logistical assistance (November 10, 1950). At first, a coalition government composed of representatives of both the Ranas and the

Nepali Congress was established, and the royal prerogatives were restored to the Shah monarch. In 1951 a Royal Proclamation was issued announcing the establishment of a new government in which no Ranas held any cabinet position. In the ensuing decade, 1951-1960, Nepal was engaged in political experimentation with constitutional Democracy. During that period twelve different cabinets were appointed including three occasions on which the monarch exercised direct rule. Evidently, no single party proved capable of mobilizing sufficient support to stay in power. Conflicts among the parties and disunity within each party led to further fragmentation of the political forces and enhanced political instability (though it did not reach the point of violent struggle). In 1959, following the first democratic elections ever held in Nepal, the Nepali Congress party emerged as the victorious political organization and started, on what was believed to be the road to Constitutional monarchy. King Mahendra, who ascended to the throne after his father's death (1955), apparently had some different ideas on the most suitable form of government for Nepal. In December 1960 a "royal coup" was staged and King Mahendra assumed direct rule. He banned political parties (January 1961) and during the next five years established himself firmly in power while inaugurating a new political system—the "Panchayat Partyless Democracy." The inclusion of the term "democracy" seems rather cynical in view of the manner in which the newly adopted system grew to resemble an absolute monarchy. Furthermore, the ban on political parties created a vacuum in the political system. This was soon filed by the administrative machinery that ever since the 1950 revolution became a tool utilized by different parties in the political arena. With the ban on political parties effectively enforced, the main competition over distribution of power and resources appeared to be taking place between the Palace and the bureaucracy as well as among the units (i.e. ministries) of which the bureaucracy was composed. In Nepal, the distinction between administrative processes and functions and their political parallels lost most of the clarity it ever had.

Nepal's geography has always strongly conditioned its history and certain salient geographic features must be recognized for adequate comprehension of Nepal's development problems, internal politics and foreign policy. Nepal is a landlocked country bounded on the east, west and south by Indian territory and on the north—along some 670 miles of the Himalayan range—by China. The physical setting is comprised of a series of parallel ranges of varying heights that traverse the country east to west. The first elevations are those of the Churia range (Siwa Lekh) that rise from the Terai lowlands. These foothills have a general elevation of 1,000 ft to 5,000 ft. Immediately to the north rises the Mahabharat Lekh range with elevations ranging from 6,000 ft to 10,000 ft. The

Mahabharat Lekh runs parallel to the Churia range and provides a formidable natural barrier to the interior parts of the country, broken by some river gorges. The third range consists of the main Himalayas situated about 60 miles north of the Mahabharat Lekh. With elevations between 16,000 to 29,000 (Mt. Everest) the Himalayan range constitute the bastion upon which South Asia relies for protection. Three principal river systems cut through the country in a generally north-south direction. Those systems, the Karnali (on the west), Gandaki (central Nepal) and Kosi (on the east) all have their sources in Tibet, and enter Nepal through gorges that bisect the Himalayas. South of the crest they are joined by numerous tributaries and eventually make their way down to the plains, where they merge with the Ganges. This river system vastly complicates east-west communication in Nepal and hampers the political and administrative unification of some parts of the country. Indeed, up until recently travellers wishing to reach points in eastern and western Nepal from Kathmandu, usually crossed the border to India and utilized its transport network to reach a point of re-entry into Nepal. Such a condition was of political and economic significance to both Nepal and India.

From a geo-economic perspective Nepal can also be divided into three distinct regions—the Terai, the Inner Terai and the Hills (inclusive of the mountains). The Terai constitutes the southernmost section of the country. It begins at the Indian-Nepali border and rises to the foothills of the Churia range. This area yields about two thirds of Nepal's total revenues. The Inner Terai, located between the Churia and the Mahabharat ranges, consists of several fairly broad valleys running from east to west. This area is currently being developed after malaria eradication programs were carried out. The Hill region, as commonly understood in Nepal, comprises the Mahabharat and Himalayas ranges and the mid-montane area between them. This latter area is the most heavily populated region in Nepal and constitutes the country's political heartland.

Having a geographical setting that kept different parts of the country isolated from one another, Nepal could not avoid assigning the establishment of transport networks as a high priority. The leadership clearly understood that without direct access to various parts of the country its ability to govern would be limited. Additionally, it was apparent that in order to attain economic growth, the Hills area and the Terai would have to be connected. It was also clear that with its limited resources, Nepal would have to depend on external assistance in order to have a transport network established.

Among the developing countries, Nepal is distinguished by the exceptionally high priority it assigned to the development of transport facilities, particularly roads. Ever since the "democratic revolution" (1951) Nepal has stressed large scale investment in the transport sector. In each of the country's development plans this sector was originally allocated about 30 percent of the total outlay. Moreover, in most of these plans the transport sector consumed approximately 40 percent of actual total expenditure| Nepal also proved highly successful in mobilizing external participation in this pattern of investment. External involvement in the form of capital and technical assistance amounted to approximately 90 percent of the total in the transport sector. External assistance in this particular sector accounted for a sizeable part of the total aid received by Nepal. Among the major external participants in this investment in Nepal were its two neighbors (India and China), three major powers (the USA, USSR and UK) and to a smaller extent, multilateral agencies such and the UN, IBRD and ADB. In Nepal, substantial investment in the basic infrastructure such as transport facilities was a necessity for economic growth. The country's geographical setting and the structure of the economy dictated the assignment of a high priority to the transport sector in every development-oriented plan. Furthermore, without establishing direct links between the center and the country as a whole, the regime's ability to govern the polity and to carry out its development-oriented policies could not be adequately exercised. Nepal's geo-political setting made it an area of strategical importance to India and China. With this external interest, not the least of which was the establishment of road networks that led to either border, Nepal could neither ignore the political significance of its transport projects nor fail to see the economic and political opportunities inherent in this particular sector. However, foreign aid participation in Nepal's development effort posed some potential problems. A variety of administrative techniques and a host of external interests could be incompatible with the internal policies and practices of the recipient. It could have serious implications on the internal political arena, on the process of development policy-making and on the implementation of development programs.

REFERENCES

1. The vast volume of literature on the various aspects of the development process cannot be overstated. Even the so-called "classical" literature (e.g. the "major contributions") has grown far too big to be listed briefly in a short introductory essay. I shall therefore list some of the major works that served as a basis for this sketch of the literature on development, politics and bureaucracy.

D. Apter, The Politics of Modernization, Chicago, 1965.

E. C. Banfield, The Moral Basis of a Backward Society, The Free Press, New York, 1958.

R. Bendix, Nation-Building and Citizenship, Doubleday and Co., Inc., Garden City, N.Y., 1969.

L. Binder, et al., Crises and Sequences in Political Development, Princeton University Press, N.J., 1971.

C. E. Black, The Dynamics of Modernization, Harper, N.Y., 1967.

N. Caiden and A. Wildawsky, Planning and Budgeting in Poor Countries, J. Wiley and sons, N.Y., 1974

K.W. Deutsch, The Nerves of Government, The Free Press, N.Y., Collier-Macmillan, London, 1966.

S.N. Eisenstadt, Modernization: Protest and Change, Prentice Hall, N.J., 1966.

A. O. Hirschman, Journeys Towards Progress: Studies of Economic Policy-Making in Latin America, The Norton Library, W. W. Norton & Co., N.Y., 1973.

R.T. Holt and J. E. Turner, The Political Basis of Economic Development: An Exploration in Comparative Political Analysis, Princeton, 1966.

S.P. Huntington, Political Order in Changing Societies, Yale University Press, Yale University, New Haven and London, 1968.

M. Landau, "Redundancy, Rationality and the Problems of Duplication and Overlap," Pub. Adm. Rev., 29, pp. 346-358, 1969.

D. Lerner, The Passing of Traditional Society: Modernizing the Middle East, The Free Press, N.Y. 1958.

C. Leys (ed.), Politics and Change in Developing Countries, Cambridge University Press, Cambridge, 1969.

J.D. Montgomery, Foreign Aid in International Politics, Prentice Hall, N.J., 1967.

J.D. Montgomery and W. J. Siffin (eds.), Approaches to Development, Politics, Administration and Change, McGraw-Hill, N.Y., 1967.

F. W. Riggs, Administration in Developing Countries, The Theory of Prismatic Society, Houghton Mifflin Co., Boston, 1964.

W.W. Rostow, The Economics of Take-Off into Sustained Growth, N.Y., 1963.

N.T. Uphoff and W.F. Ilchman, The Political Economy of Development, Theoretical and Empirical Contributions, University of California Press, Berkeley, Ca., 1972.

E. Weidner (ed.), Development Administration in Asia, Duke University Press, Durham, 1970.

M. Weiner (ed.), Modernization: The Dynamics of Growth, 1966.

2. N. Caiden and A. Wildavsky, Planning and Budgeting in Poor Countries, op. cit., p. 31.

3. For a discussion of some major aspects of transportation in the development effort, see W. Owen, Strategy for Mobility, The Brookings Institution, Washington, D. C., 1969.

4. See, for example, G. Fromm (ed.), Transport Investment and Economic Development, The Brookings Institution, Washington, D.C., 1965. Numerous articles approaching the matter from either economic standpoint or engineering standpoint appeared in professional journals of those disciplines as well as in area studies journals.

5. This data on Nepal's economy is based on publications by Nepali government agencies, the World Bank and the United Nations.

6. For literature on Nepal's history, political system and economy, consult the references at the end of Chapter 3.

Chapter 2

TRANSPORTATION AND DEVELOPMENT IN NEPAL: PRIORITIES, PLANS, PROJECTS, 1951-1975

Since 1951 Nepal's development programs have been dominated by large investment in the transport sector, in particular road construction projects. The country which up until the early 1950s was a land with almost no modern transport facilities, possessed by 1976 and 3200 km road network. This achievement was possible mainly because several aid-donor countries were interested in investing in this particular sector of the Nepali economy. Nepal's central regime had its own reasons for emphasizing this particular sector. The following discussion reviews in detail Nepal's development efforts in the transport sector during the years 1951-1975. It discusses priorities and plans set for the transport sector and analyzes the planning activities undertaken in this field in Nepal. The review presents the characteristic features of all major road projects carried out in the country. It analyses the pattern of budgetary allocations and examines the sources of financial support as well as the progress attained in road construction operations in each of Nepal's four successive development plans.

Transport Facilities in Nepal 1951-1956

The Rana regime, pursuing a policy of isolation from the outside world and maintaining its control over the country through familial and group networks, had neither wished nor tried to establish an elaborate communication system in Nepal. Its domestic policy was based on this maintenance of the status quo and, with the exception of agriculture, it did not enhance progress in the economy. In 1951, when the Ranas were toppled Nepal was "a land isolated not only from the outside world but also within itself, one part from the other . . . when the curtain of isolation was torn aside . . . it revealed a country without a transport system or communication network, illiterate, backward and poverty ridden."[1]

Indeed, whatever transport facilities existed in Nepal at the time were mostly centuries-old means of transportation—primarily porters and, to a smaller extent pack animals and bullock carts. Most traffic moved on mountain trails and hill footpaths. In 1951 the total length of roads in the country was 376 kilometers—6 kilometers of pitch road, 80 kilometers of gravelled road and approximately 290 kilometers of fair-weather roads. This meager roads network was supplemented by a few miles of railway and some air transport facilities in Kathmandu, Simra, Bhairawa and Pokhara (of which only the first was an all-weather airport).

The two railways which existed at that time were the Nepal Government Railway (NGR) and the Nepal Jaynagar-Janakpur Railway (NJJR). The NGR served as a major avenue of commerce into and out of Kathmandu Valley. It was constructed in 1927 and ran for 43 kilometers from India to Birgunj and Amlekhganj. Although poorly maintained it remained profitable up until 1956 and provided both passenger services and goods-transportation service.[2] The second railway, the NJJR was constructed in 1937-8 connecting Jaynagar, India, with Janakpur. This 27-kilometers-long line was originally built to move timber from the Terai, just above Janakpur, to the Indian railway at Jaynagar. Once all the timber in the area was cut, the line became a means of transport for the religious pilgrims who came from India to the Hindu shrine in Janakpur.[3] Both lines are still in operation. The NGR is used for the transportation of goods, whereas the NJJR, closed during the 1960-66 period, now provides a passenger service in the Janakpur area.[4]

Kathmandu, the capital, and the relatively more developed area in the country, was served by yet another means of transport—the ropeway. This was the only means of transporting freight into or out of the Kathmandu Valley other than by porters. Constructed in 1927 the ropeway line stretched from Bhimphedi to Kathmandu. Its terminal in Bhimphedi was located at the end of a narrow but truckable road leading from Amlekhganj. Thus, at least insofar as freight transport was concerned, both this ropeway and the NGR constituted the major links between Kathmandu and the Indian border. Suffering from poor management and lack of proper maintenance the ropeway service deteriorated after 1951.[5] With the opening of an alternative road link, the Tribhuvan Rajpath (highway) in 1956, it lost most of its customers. This ropeway line was finally replaced by a new one in 1964.

The use of waterways within Nepal was at best very limited. Existing rivers, particularly the Gandak, were used for some movement of rice paddy and merchandise to and from India and for floating logs out of the hill forests to the Terai. Other than that there were no other navigable rivers or canals in the country.

In 1951 Nepal had no transport system capable of meeting the needs of an economy that seeks to grow and develop. It became apparent to all that the introduction of modern transport facilities, substituting for the traditional means in use, had to be the prime objective and an essential precondition for growth in other sectors of the economy. It was also recognized that due to the acute scarcity of resources, Nepal could not undertake this alone.

During the years 1952-1956 two of the major development projects carried out in the country were in the transport sector. Shortly after the Shah dynasty resumed power, Nepal accepted an Indian offer for the construction of a land-route linking Kathmandu with the Indian border. The road was constructed by the Indian Army from 1952-1956 and led from Bhainse to Kathmandu. it was the first motorable road linking the capital with the outside world. During that period the British Army embarked on the construction of another road further east. This second road was to connect Jobani on the Indian border with Dharan. Once completed it provided a direct link between the Indian railhead opposite Biratnagar and a Gorkha recruitment camp in the foothills to the north. Some additional kilometers of inner-city roads were also built during that period primarily in Kathmandu and its immediate surroundings. Though these projects doubled the total length of the existing roads in Nepal, they were not the outcome of a plan formulated by Nepali authorities. The Indian-built road was constructed to serve India's own interests in the area at least as much as it was meant to be a neighborly contribution to Nepal's development effort. Similarly, the launching of the British project was based primarily on the donor's interest in obtaining better access to a Gurkha recruitment camp for the British Army. Lacking the necessary capabilities and resources to carry out those undertakings, Nepal had to accommodate the interests of the aid donors in order to have those roads constructed. The case of the Dharan road illustrates this point well, since this area had not been given high priority for development by Nepali authorities.

Development Plans: Priorities and the Transport Sector

The establishment of a transport network on a national basis became one of the prime objectives for the newly reinstated Shah regime. The central authorities recognized the importance of such a network not only as a basic precondition for development in other sectors of the economy by also for political and administrative considerations. The immediate past provided one crucial example for the political importance of transport facilities in Nepal. As some observers suggested, the refusal to establish transport and communication links in the country proved to be the Achille's heel of the Rana regime:

> For in Nepal's geographic context the cost of not investing in a national transport system was dependence on the Indian railways even for movement from one part of Nepal to another. So, when the rebel movement went into action in 1950-51 with implicit Indian support, Nepal's total dependence on the Indian railway system

> proved fatal to the Ranas. India did not have to provide significant military support to the rebel movement. She merely denied the use of the railways to the Rana troops while permitting its use to the rebel movement. This one fact, the advantage of movement, accounts for the early and dramatic success of the "progressive" forces in major towns.[6]

In 1960-62, during and immediately after the "royal coup" the issue of transport facilities once again became strongly entangled with political developments; at that time the king proved to be fully aware of the importance of transport facilities for military operations and succeeded to suppress the rebels; India, although supporting the "democratic forces," acted much more carefully where provision for full transport facilities for the rebels was concerned.

The first Five Year Plan* pointed out that the lack of a transport and communication system was an obstacle to all development plans:

> People living in small and isolated rural communities are cut off from each other. Each isolated region tends to follow a pattern of subsistence economy of the most primitive and rudimentary type . . . administration is costly and the government cannot fulfill effectively its role of promoting the people's welfare; the existing exploitation of resources is inefficient and potential resources cannot be developed.[7]

The call therefore was made for both the rapid construction of a system of moderately priced roads which would give the country year-round service for its most pressing business needs and the collection of technical data necessary for projected plans and future projects. Indeed, of the total estimated budget for this plan's period (1956-1960) the transport sector was allocated 31 percent. By the end of the First Plan period the sector's share of total actual expenditure had amounted to 41 percent.

In the Second Plan (1962-1965) a somewhat different set of priorities was established for the transport sector. Using practically the same rationale to emphasize the importance of this sector, the

---

*This plan, titled "Draft Five Year Plan" was mainly an outline presenting the major needs of the economy in broad general terms. No comprehensive plan or detailed order of priorities could be prepared due to lack of essential statistical information.

plan allocated 39 percent of the total estimated outlay to communications.[8] However, unlike the previous plan more emphasis was put on the development of electric power facilities and aviation. Consequently, the estimated allocation to development of transport facilities was reduced to 23 percent of the plan's total outlay. Furthermore, by the end of the Second Plan this sector's share of total actual expenditure was 14 percent. This decline is explained not so much by reduced importance assigned to this sector, but by the fact that funds for certain projects were not included in the budget. Moreover, it was only toward the end of the plan period that major road projects were started, and the actual expenditures on these were either small or transferred to the next plan's budget. Both the First and the Second Plans were hastily prepared, and neither was a comprehensive, well-coordinated and implementable program for development. Both were prepared under conditions of insufficient data and lack of overall approach toward the economy. Consequently, both were more or less broad lists of expenditures in the public sector. In fact, they could not be considered more than a preparatory effort to create the basis for a comprehensive nationwide plan. Such a basis was apparently in the making during the preparation of the Third Plan.

Unlike the preceding plans, the Third Plan (1965-1970) was formulated on the basis of a 15 years planning perspective. It was broader in scope and included not only the public sector but the local panchayats and the private sector as well. It envisaged overall investment outlays averaging Rs 500 million annually (as compared with Rs 600 million during the whole Second Plan period). And, apparently, it was based on more accurate data than previous plans. The increased volume of data was mainly the result of information gathered by a World Bank team that was invited to Nepal to produce a Five Year Transport Sector Plan for inclusion in the Third Five Year Plan. In accordance with a policy of large investments in infrastructure, the transport sector in the Third Plan was allocated 35 percent of the estimated total outlay.[9] By the time this plan's period ended, the transport sector's share rose to 37 percent of total actual expenditure.

This pattern of assigning high priority to the development of transport facilities continued in the Fourth Plan (1970-1975) that emphasized:

> It is only through transport expansion that development projects of all types can flourish together through mutual coordination and integration . . . it is evident from our experience (as well as other countries) that unless transportation is sufficiently developed, development programmes cannot be effectively implemented. Consequently, transportation has been given top priority.

MAP 1

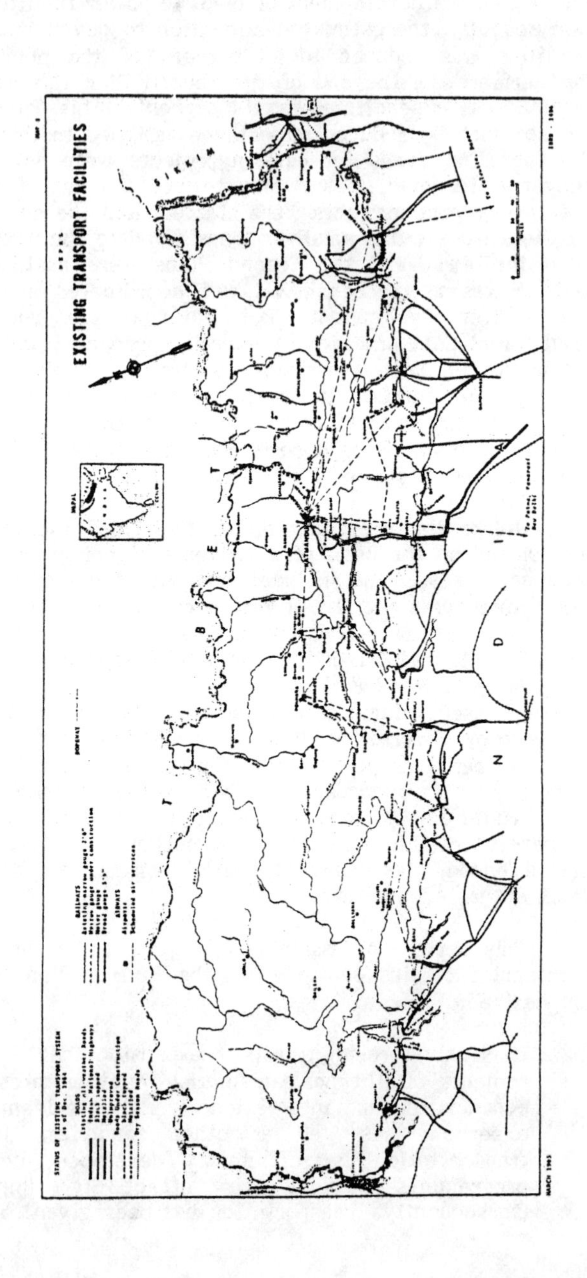
EXISTING TRANSPORT FACILITIES
SIKKIM
TIBET
INDIA
KATHMANDU

> Furthermore, its development is necessary to bring about efficiency and coordination in administration, to overcome regional disparity and to mobilize scattered and surplus labour from agriculture to other sectors.[10]

Thus, in the Fourth Plan the transport sector was allocated 39 percent of the estimated total outlay. During the Fourth Plan, development programs in the transport sector were coordinated on a more national basis and according to a long-term plan. By the end of the Fourth Plan period the transport sector's share of total actual expenditure amounted to 40 percent.

Planning a National Transport System for Nepal

Until 1965 Nepal did not have a comprehensive, nationwide plan for development. Priorities were determined more on the basis of availability of external aid rather than on the basis of a sound long-term plan for development. A general set of priorities did exist, but it was based on rudimentary ideas that were subject to interpretation in too many ways to be a useful guideline for action or coordinated effort. Nevertheless, in view of the central role that the transport sector came to occupy in the country's development activities, the regime appeared to consider it the most suitable field in which long term planning for development could succeed and subsequently provide the necessary impetus for planning in other sectors of the economy. Accordingly, Nepal requested the World Bank to assist her in the preparation of development plans.

In 1964 a World Bank's Nepal Transport Mission was assigned to prepare a Five Year Transport Sector Plan, for inclusion in the government's Third Five Year Plan. The Mission, "due to lack of clear statement of the national goals for other key sectors of the economy and in view of serious shortage of basic economic and statistical informations of all kinds," found it necessary to approach the subject matter on a much broader basis than originally planned.[11] It gathered all available background information, assembled and analyzed it. On the basis of this data a long-term Master Plan for Transport Development was drafted for the purpose of establishing specific objectives and a framework for future planning. A specific plan for the Third Five Year Plan was then abstracted from this Master Plan.*

---

*The "spin-off" from gathering this information and assembling the data was the substantial increase in the volume of data available on Nepal's economy. Thus, a basis for further planning exercises in other sectors of the economy could be started and included in the country's development plans. Such a phenomenon might be

In 1965 the Mission submitted its three-volume report.[12] In 1966 the Roads Department, HMG, assisted by UNDP as the executing agency, and on the basis of the Master Plan, prepared a perspective Road Plan and expanded it into a Revised Twenty Year Road Plan (published in 1967).[13] This plan "identified the nodes to be connected by road, designed the general shape of the network, assessed the length of roads needed and, estimated costs on an 'order of cost' basis. General financial assessments were made for construction, maintenance, and establishment costs and an annual expenditure forecast over the twenty year period 1965-1984."[14]

The plan anticipated that the bulk of construction finance would come from bilateral and multilateral agencies in the form of grants and loans. This plan is now being gradually implemented. Some recommendations were included in the Third Plan, others in the Fourth and Fifth Plans. Several more recommendations were ignored by each of these plans. In view of these facts and in order to provide the necessary background on developments in this sector since 1965, the following discussion presents the main principles of the Master Plan.

The Master Plan[15]

The overall plan calls for the development of a basic transportation network designed to serve the principal needs of the country over a long run. It combines a skeletal system of national highways with a program for the development of a commercial air transport network and some small trans-border terminal railways. The plan calls for new construction plus improved maintenance and operation of existing facilities, including the ropeway. This system of national highways in the Master Plan is to be supplemented by a system of local and feeder roads.

The principal recommendations were:

(1) Construction of a 2400 mile national highway system reaching all of the more heavily populated regions of the country;

(Continued)
considered an example of the case where even though skilled personnel is available in the recipient country, it would not be able—due to budgetary, administrative, and possibly political constraints—to start such a project on its own. However, once the initial step is made by an external, "neutral" donor, the basis for further action is there for local personnel to use.

(2) Providing the means for the creation of a system of "Panchayat roads"—generally local and feeder roads built and maintained by the local Panchayats, linking with the primary network;

(3) Establishing a system of ten commercial airports with paved runways and designed for year-round service by Royal Nepal Airlines Corporations (RNAC);

(4) A reorganized RNAC equipped with better planes and designed to operate as a self-supporting commercial airline;

(5) Provision of three cross-border meter-gauge terminal railways closely tied to the Indian railways and designed to simplify customs and reduce handling costs;

(6) Rehabilitation of the NJJR narrow-gauge cross-border line from Jaynagar to Janakpur;

(7) Improved operation and management of the Kathmandu ropeway;

(8) Creating a system of automobile and truck servicing and maintenance depots.

The plan estimated that all these elements could be built within a period of approximately 30 years, provided that the current trend of investment in the transport sector continued.[16]

The Road Transport Plan

In this overall plan, roads and road transport were selected as the prime transport media in Nepal. The reasons for the selection were indicated by the Mission as follows: First, roads bring the largest amount of economic development in relation to the total expense. The Mission apparently concluded that although the investments needed for construction of highways would be substantially larger than for most other forms of transport facilities, the projected needs and capabilities of the Nepali economy would be best served by this form of transport technology. The Mission referred here mainly to the potential benefits that would accrue during the construction period—e.g. benefits related to new employment opportunities, income distribution, vocational training and the emergence of new markets along the alignment of the roads under construction. Second, over a long span of time roads provide the lowest cost total transportation. Third, roads can be used in many ways for many different purposes, an implicit reference to political, strategical and administrative considerations involved in the establishment of a road network. Accordingly, the plan called for establishing a classification of roads, including a distinction between national highways and Panchayat roads; strengthening the Roads Department (to be called the Road Authority) for better handling of maintenance and construction work; mobilization of the Panchayats for road construction and maintenance; and development of a national system of all-weather roads.

The development of a national highway system, it was emphasized, must necessarily be spread over a long period of time due to the high level of investment involved. Yet,the Mission also pointed out that the rationalization of road construction and road maintenance was a timely and critical problem in need of immediate and continued attention.[17]

Classification of Roads

The Mission advised that there be two distinct systems of roads in Nepal. The first would be heavily-used, all-weather roads called the National Highway System (NHS) that would constitute the skeleton on which local systems of dry-season roads and foot-tracks could be based. It was further emphasized that although the NHS must serve all areas with heavy population density, it should be kept as short as possible. The second system of roads would include all other classified roads (i.e. local roads) outside the NHS. For purposes of construction and maintenance work, the Mission recommended that a centralized, well-equipped and well-staffed organization (to be called the Road Authority) be placed in charge of the NHS while the Panchayats would be given responsibility for the second system.[18]

General Pattern of the National Highway System

In preparing the general pattern of the NHS the Mission considered the country's topographical setting, locations of broad population concentrations, projected costs and expected contribution to development. It described the circumstances as follows:

> There are two broad population concentrations in Nepal—one along the Terai, and the other in the hills somewhat back from the Terai. Between the two there is a zone of forest with poor and unstable land, which is lightly populated. The largest volumes of trade and the most important traffic is generally north-south.

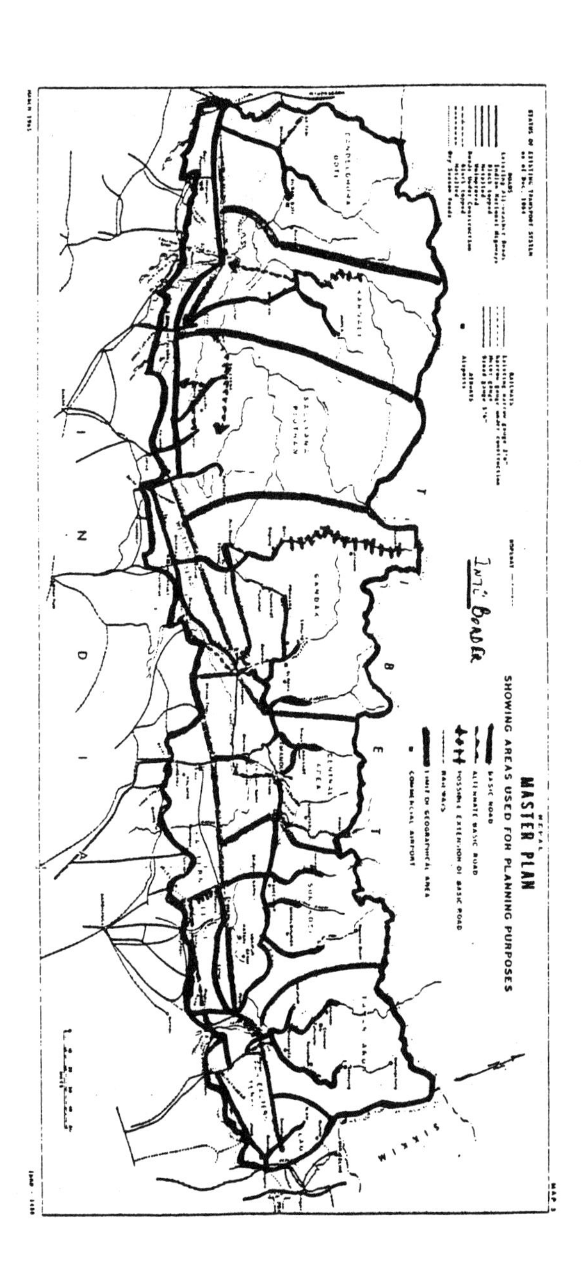
MASTER PLAN
SHOWING AREAS USED FOR PLANNING PURPOSES
INT'L BORDER
GANDAK
SIKKIM

MAP 2

> However, because the country is a long east-west strip, there is a need to provide some internal linkages between east and west.[19]

The Mission suggested therefore that the best general strategy to use is to follow principal river valleys. In this way construction, maintenance and transport costs would all be reduced. It was further pointed out that:

> In the Terai, north-south routes present few problems and are generally easy and economical to construct, because they follow the local watersheds and thereby generally avoid wet ground and river crossings. More troublesome is the problem of east to west linkages along the Terai—the most obvious route for running the length of the country. From a purely technical and cost point of view, the best route for such linkage is the northern edge of the Terai, at the base of the foothills. Here river crossings are easier and construction materials (principally gravel) are in relatively abundant supply. However, such a route completely bypasses all principal areas of population and of agricultural activity. These are generally located further south in the Terai, near the Indian border.

Faced with this situation the Mission concluded that even though the costs of a road serving those areas to the south would be higher, greater economic benefits could be obtained from it.

Of particular importance in the overall plan was Nepal's largest transportation project—the East-West Highway. First proposed by King Mahendra in 1961-62, the project was meant to serve not only development needs but also (or, as some have argued—principally) to symbolize the country's growing independence. Furthermore, HMG expected it to be the "spinal cord" of road system in Nepal. The Mission, in preparing the general pattern of the NHS had to consider it as given even though work on this project was not yet started. The Mission therefore commented:

> A highway linkage from east to west is necessary from an administrative point of view. However, this alone is not sufficient justification for building an improved road. Therefore the principal justification for such a route must rest on other values for each of the individual sections. This would include access to markets, movement of commercial and export crops, improved agricultural development etc. For these reasons the Mission considers it necessary to design these roads so

Figure 1

PRIMARY ROAD NETWORK—NEPAL
(Diagrammatic)

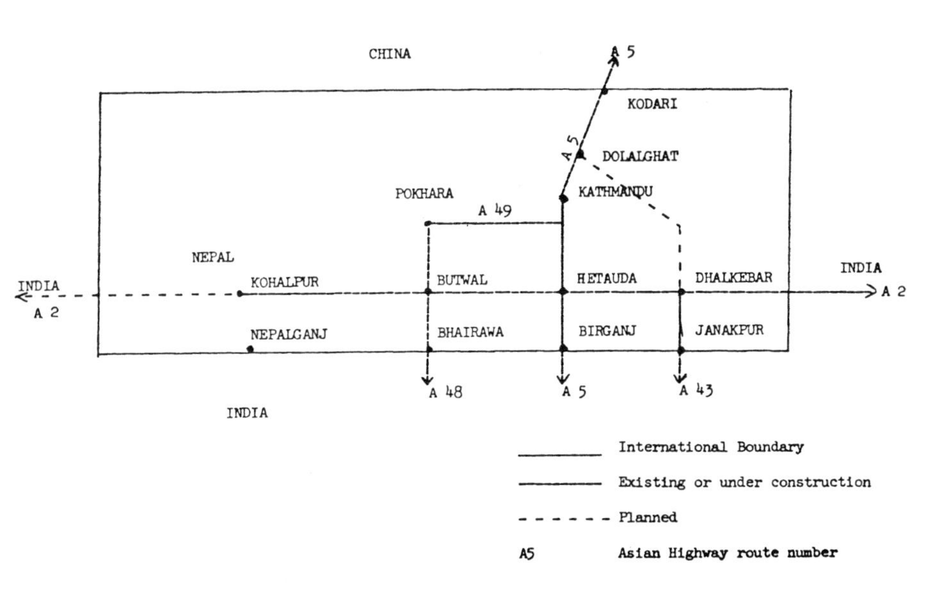

they will be of maximum local utility and only incidentally provide an east to west linkage.[20]

On the whole the Mission recommended a 2700-mile road network for Nepal. Of this total, 2400 miles were to be a first priority network while the other 300 miles would constitute second priority northern linkages to Tibet.[21] Essentially, the primary road network of Nepal was to consist of an east-west spine road (Asian Highway Route A-2), supplemented by a central rectangular box. (Asian Highway Routes A-48, A-49 and A-5) that would provide closer integration of the rest of the country with Kathmandu (see attached diagrammatic representation and Master Plan Map).

Planning the Secondary Road System

By early 1968 the form of the primary (trunk) network was well established, with large proportions either completed, under construction or planned for construction by various agreement between HMG and bilateral agencies.[22] However, it became apparent that the investment in the primary road system could not be fully exploited unless the secondary system was also constructed in phases. The secondary road-network therefore was recognized an essential part of the entire development scheme in the transport sector. Moreover, the Twenty Year Road Plan recommended that planning and feasibility studies of a secondary road system be initiated as an essential first step in the implementation process of the whole program. This was in congruence with the government's declared objective to integrate the Hills and the Terai, economically as well as socially, and to provide balanced development throughout the country. Such objectives, translated into the terms of secondary road development broadly meant designing a network that would penetrate northward and southward from the primary road system; a network that would connect all district and zonal headquarters with the primary system and through it, with each other and the center at Kathmandu. The location of the secondary network would be governed by both the north-south alignment of the river valleys draining through the hills from the Himalaya to the Gangetic plain and, by the location of the major towns and markets in the central hills belt where two-thirds of Nepal's population live.

In September 1968 HMG requested the assistance of UNDP to plan the secondary road system. The project was approved in early 1969 and actual work started in June 1970.[23] Its objectives were as follows: (1) reconnaissance surveys to determine the need for secondary roads in Nepal, to identify the form and shape of the network that would satisfy the needs and to select from the total network an initial package of 1000 km of roads for feasibility study as a preliminary step to construction; (2) feasibility studies to carry out economic and feasibility studies of the 1000 km so selected; (3) preliminary Investigations to study the present transport situation in the Kathmandu-Birganj corridor,[24] and to make recommendations as to what steps HMG should take to provide adequate transport facilities in the future; and (4) in 1972, upon completion of objective 3, HMG requested further assistance in carrying out technical and economic feasibility studies of two alternative new roads between Kathmandu and Birganj.[25]

These four objectives were subsequently the subject of separate reports by the consultants.[26]

The secondary road network ultimately identified by the Reconnaissance Survey, consisted of 140 links totalling approximately 5000 kilometers. A package of eighteen roads, with an estimated total length of 950 kilometers, was then selected for further study. The selection of this package was based on the evaluation of the ratio between the overall economic weight of the districts served and the construction and maintenance costs of the roads, together with an assessment of the increased accessibility the new link would provide.[27] During detailed economic and technical surveys it became apparent that modifications to routes and, in some cases, the end point of the road, would provide better economic service to the area concerned. In other cases technical problems, mainly concerned with geology and slope stability, entailed some rerouting of road lines. The combined result of these economic and technical relocations was an increase of the length studies to 1116 kilometers.[28] On the basis of that latter study a construction program covering the 1975-1985 period was recommended for thirteen roads that were found economically viable. Additional findings and recommendations on these roads and steps to be taken during implementation were furnished in a 1975 UNDP report.[29] Implementation was expected to follow after those findings and recommendations were reviewed by HMG.

By 1975 Nepal had a comprehensive plan for the transport sector and, a lot of mileage to cover.

Major Transportation Projects—Roads

In Nepal, external agencies carried out all the major road projects, most of which were executed on the basis of bilateral agreements between HMG and different aid-donor countries—primarily India, China (PRC), USA, USSR and UK. Several multilateral agencies like the UNDP, IBRD and ADB were also involved in some of the projects. The following discussion reviews the major road projects which were carried out in Nepal during the last twenty-five years. This review presents the main characteristics of each project—its location, projected contribution to development, time of construction, executing agency and some technical-technological features involved in it. The political and administrative aspects of these projects will be discussed in the next chapters.

Tribhuvan Highway (Kathmandu-Raxaul)

This highway was the first to be constructed in Nepal and the first to provide a link between the capital and the outside world by means of a land route. The road, built by the Indian Army during 1953-1956, is 191 kilometers long and named after King Tribhuvan. It constitutes the major trade link between Nepal and India. Starting from Kathmandu it leads in a southwardly direction to Thankot, Nagdhunga, Naubise, Tistung, Palung, Daman, Simbhanjyang (8162 feet), Bhainse, Hetaura, Amlekhganj, Pathlaiya, Simra, Birganj, and to the Indian border town of Raxaul. The road, so constructed, connects the Bagmati zone with the Narayani zone, both in the Central Development Region.

Once completed in 1956, this highway replaced both the old railway (NGR) and the ropeway in providing service to most passenger transportation and transport of goods. It is by far the most heavily utilized highway in the country and, for the flow of goods, undoubtedly the most important. However, heavily used and inadequately maintained, the road has been described as "extremely tortuous."

> It is well endowed with loops and tight curves, one-lane bridges, narrowness of carriageways, poor quality pavement and high rate of of rise and fall. The traffic density is such that on two-thirds of its length it is operating under congested conditions, resulting in rising user costs and an increasing accident rate.[30]

The apparent deficiencies of that alignment were in fact recognized as early as 1960-1961, but the proposals for an alternative road proved unfeasible. In 1968 HMG submitted a formal request to UNDP for assistance in carrying out Road Feasibility Studies. Among the project requested was a preliminary investigation of the transport situation in the Kathmandu-Raxaul corridor. The request was approved in 1969 and by 1971 the preliminary investigation was completed. It recommended two alternative new roads between Kathmandu and Birganj (Raxaul) on which further economic and technical studies would have to be conducted.[31] In 1972 additional assistance was approved for the purpose of carrying out those studies and by 1974 recommendations on the more advantageous road of the two alternatives were submitted to HMG. Pending approval of those recommendations, detailed design work was expected to follow, with construction work planned to start in late 1977. The Tribhuvan Highway is part of the Asian Highway Priority Route "A-5."

The Arniko Highway (Kathmandu-Kodari)

This highway, built by the Chinese during the period 1962-1967, is 114 kilometers long and is the only major road that connects Nepal with the territory of the People's Republic of China. Agreement on the construction of this road was reached, rather unexpectedly in October 1961 during King Mahendra's visit to Peking in which a new boundary agreement between the two countries was signed.[32] The road is potentially of substantial strategic importance since it offers one of the easiest routes through the entire Himalayan range into the India sub-continent. Starting from Kathmandu the Arniko Highway leads in a north-northeast direction to Bhaktapur, Banepa, Dhulikhel, Panchkhal, Dolaghat, Balephi, Barabise, Tatopani, terminating at "Friendship Bridge" near Kodari. This road is rather lightly utilized and, though it did open up some previously inaccessible areas in that part of the country, its economic impact has been rather limited. The Arniko Highway also constitutes part of the Asian Highway Priority Route "A-5" and, together with the Tribhuvan Highway, provides a land route that crosses Nepal north-south from Tibet to the Indian border.

Sidhartha Highway (Pokhara-Sunauli)

A 184-kilometer highway built by India in 1964-1972, the Sidhartha Rajmarg links Pokhara—the second largest valley in Nepal--with the Indian town of Sunauli. Starting from Pokhara, in the Western Development Region, the road stretches in a southwardly direction to Nuwakot, Syangja, Tansen, Ranidighat, Bartung, Butwal, Bhairawa and Sunauli. The Sidhartha Highway thus provides a link between the Western Hills and the Western Terai as well as a land route that connects Inner Nepal with the Indian border. The opening of the road served to concentrate the flow of goods north into Nepal. Yet, the extent to which it succeeded in inducing economic growth remains quite limited.[33] The road is of some cultural-religious importance because it connects Buddha's birthplace (Lumbini) with Pokhara. The road is also known as Route "A-48" of the Asian Highway System.

Prithvi Highway (Kathmandu (Naubise-Pokhara)

The Prithvi Rajmarg, built by China during 1966-1973, provides a land route link between Kathamndu and Pokhara. This 174 kilometer road thus connects Nepal's Central Development Region (Bagmati) with the Western Development Region (Gandaki). Starting from Naubise (on the Tribhuvan Highway) it runs in a westwardly direction to Dharke, Gajuritar, Benighat, Charaudi, Muglingphant, Damauli and Pokhara. The distance from Kathmandu to Naubise is 26 kilometers on the Tribhuvan Highway. The Prithvi Highway can be considered the counterpart in the Hills to the Terai's road (Mahendra Highway Hetaura-Narayangha-Butwal portion).[34] Consequently this Western Region of Nepal is the only region that has two parallel major highways. Economically the road's most significant contribution is that of providing opportunities to tourists visiting Kathmandu to travel further west into Nepal. Administratively it provides an important link between the center in Kathmandu and the Western Region.

Mahendra Highway--The East-West Highway

This project, the largest in Nepal, involves the construction of 1040 kilometer highway which, once completed will cross the country from east to west through the Terai Plains and the Inner Terai. The call for the construction of this highway was made by King Mahendra in 1962. It was meant to provide Nepal not only with its first cross-country land route but also with the transportation facilities that would obviate the necessity of passing through Indian territory on east-west journeys. It was assigned topmost priority to the extent that, despite initial reluctance by all major donor countries to participate in it, they all became involved in the construction of its various portions.[35] It is often referred to as the "spinal-cord" of Nepal's road system; moreover, for many Nepalis, it symbolizes Nepal's independent status.[36]

The Far Eastern Portion (Kankarbhitta-Dhalkewar)

This portion of the Mahendra Highway, completed in 1976, was under construction since 1966. It extends from Mechi on Nepal's eastern border, to Dhalkewar close to the Central Development Region (where it links up with the "Russian" section of the highway). The road connects the Mechi, Kosi, and Janakpur zones and is expected to increase the volume of internal trade. This road was constructed by India which initially showed no interest in the project, but agreed to construct it after China expressed its willingness to do so. India reportedly did not wish the Chinese to be working so close to its border.[37]

During the Fourth Plan the original agreement was further extended to include the construction of the KAMLA bridge. Evidently, it became apparent that without this bridge which connects two parts of the road, the utility of the entire highway would be greatly reduced. Moreover, India also undertook to construct a link road from Dhalkewar to Jaleswar via Janakpur. Thus, the Mahendra Highway was also connected with the Indian border.

The Eastern-Central Portion (Dhalkewar-Pathlaiya)

The eastern-central portion of the highway, also known as the Simra-Janakpur Highway, was constructed by the USSR during 1965-1973. The Russians, who conducted the initial survey on the East-West Highway, undertook the construction of this portion after it became clear that other aid-donor countries intended to take part in the project. This portion of the highway is 110 kilometers long and connects the Janakpur zone with the Narayani zone. The road passes mostly through the Terai jungle and crosses the Bagmati and Ratu rivers; it has 24 major bridges, 68 medium and small bridges, 74 culverts and 23 causeways. Unlike most other portions of the highway, the construction technology adopted in this sector was highly mechanized.[38] The road links up with the Kathmandu-Raxaul corridor at Pathlaiya.

The Central Portion (Pathlaiya-Hetaura-Narayangha)

This portion of the Mahendra Highway stretches for 110 kilometers across Nepal's Central Development Region. Both the Pathlaiya-Hetaura section and the Hetaura-Narayangha sections existed before the Master Plan for Transportation came into being. As far back as 1955 the USA built an 83 kilometer fair-weather road from Hetaura to Narayangha as part of the Rapti Valley Project; during the First Five Year Plan period both sections were gravelled to become all-weather roads. Currently this road is being reconstructed by HMG to upgrade it to highway standard. The project is of particular importance in this area since it complements two other development projects there—the Chitwan Valley Development Project and the Gandak-Hetaura Power Project. The road therefore became part of a comprehensive development program.

In addition to the reconstruction of the road the project also includes construction of a two-lane bridge over the Narayani River.

Once completed the bridge would replace the cable-ferry that has posed a serious transport bottleneck and has been found to be inadequate in view of the projected completion of the Narayangha-Butwal section of the highway.[39] Both the road and the bridge projects are financed by a loan from the Asian Development Bank, and the work is being carried out by the National Construction Company of Nepal (NCCN). This is one of the few major road projects executed by a Nepali agency.

The Central-Western Portion (Narayangha-Butwal)

This portion of the Mahendra Highway, planned as a fair-weather road during the Third Plan period, was completed in 1976 as a 120 kilometer all-weather road. Built by the UK, the road links the Narayani and Lumbini zones in the Central-Western Development Regions. It was expected that by connecting Narayangha and Butwal (on the Sidhartha Highway) increased economic activity would result in the area which had thus far been inaccessible.[40]

The Western Portion (Butwal-Kohalpur (Nepalgunj))

Construction of this portion of the Mahendra Highway was approved during the early part of the Fourth Plan and India assumed the responsibility to build it. Work on this proposed 251 kilometer long road has begun, but it is progressing slowly. This section of the highway is expected to connect the Lumbini and Rapti zones, thus linking the various districts in the western Terai. Furthermore, upon completion it is expected to link up at Kohalpur with the proposed south-north road leading from Nepalgunj to the far western headquarters in Surkhet. It would thus establish a network that links the Western Hills with the relatively richer areas in the Western Terai.

Far Western (Kohalpur-Bhanbasa)

The 200 kilometer far western portion of the Mahendra Highway is the last section of the East-West Highway. It is expected to stretch across the Far Western development Region to the Indian border near Bhanbasa. HMG took it upon itself to construct this portion of the highway which would link the Bheri, Seti and Mahkali zones. Up until now no serious steps have been taken toward the implementation of the proposed project. Once underway and completed this portion is expected to link up with with south-north road built from Dhangari to Dandeldhura.

Danghari-Dandeldhura

The Dhangari-Dandeldhura road, was the first major road in the Far Western Development Region. As early as the mid-1950s proposals were made for the construction of a road from the Far Western Terai into the Hills but none materialized. In 1960 the Regional Transport Organization began actual construction on the proposed alignment.* It conducted a reconnaissance survey and completed some earthworks as well as a two-foot track up to kilometer 48. In 1962 the RTO was dissolved and consequently the project was discontinued.

In 1968/9 the project was resumed as a joint American-Nepali undertaking (75 percent financed by the US and 25 percent by Nepal).[41] Its completion has been hampered by numerous problems due to the project's remote location and the mountainous terrain through which it passes (120 kilometers).[42]

Consequently the completion date, set at first for 1973 was later moved to 1976 and then reset for 1979-80. However, this project was considered of particular importance to Nepal since it was the only major undertaking in the transport sector that provided Nepali engineers, supervisors and technicians with actual responsibilities and direct roles in all stages of the work. Although the larger part of the project was financed by USAID, direct administrative responsibility for the construction and maintenance of the road rested with Nepal's Roads Department. The USAID furnished advisory services, mostly on technical matters.[43]

Other Road Projects

Jogbani-Dharan

This 53 kilometer road was built by the British Army during the 1953-1958 period. Located in eastern Nepal this road was constructed to provide a direct link between the Indian railhead in Jogbani and a Gorkha recruitment camp in the foothills to the north (Dharan). It was meant to serve the British Army needs at least as much as it was meant to contribute to Nepal's development effort. The road was among the first motorable roads built in Nepal and still provides an important link between the eastern Terai

---

*Formed in 1959 the RTO was a joint undertaking by Nepal, India and the USA; it main purpose was the construction of some 900 km of roads in Nepal, primarily from the Terai into the Hills.

and the Hills. Moreover, plans are now being made to extend the road further north from Dharan to Dhankuta.

Nepalgung-Surkhet

The Nepalgung-Surkhet road, which links the western Terai with the Hills, was proposed in Nepal's Master Plan for Transportation System.[44] During 1970-1973 a feasibility study on such a road was carried out in the area but no actual steps were taken for its construction.[45] In 1976 a major change in the planning strategy for the region took place, designating Surkhet as the Development Center for the Far Western Development Region. Consequently, an 81 kilometer road that connects Nepalgung in the Terai with Surkhet in the Hills received the highest priority for construction. Following a Royal Directive, resources were diverted to this project and construction work picked up momentum. The road, considered at first as a potential Nepal-USA joint undertaking became a Nepal project constructed with HMG's own resources.[46] It leads in a south-north direction and en route links up with the East-West Highway at Kohalpur.

Kathmandu-Trisuli

The 69 kilometer Kathmandu-Trisuli road leads from Kathmandu in a north-northwest direction to Trisuli. The road was built by the RTO though most of the work was carried out by India. Completed in 1963, it provides a land route to the hydroelectric dam in Trisuli (another Indian project). Furthermore, the opening of the road provided better access to the Mustang area which India considered strategically important for its own security.

Kathmandu Ring Road

A 27 kilometer "Kathmandu Turnpike"; the Kathmandu Ring Road is intended to provide better transport facilities in and around the Kathmandu Valley and relieve some of the traffic congestion in the capital. The road is being built by the Chinese and is nearing completion. Starting from Sinamugal it passes through Pashupati, Gaushala, Chabahil, Dhumbarhi, Maharag-gung, Congabu, Balaju, Swayambu, Dalimati, Balkhu, Jawalakhel, Lalitpur and Koteswar. The immediate needs for such a road are somewhat questionable and the cost of expropriating the scarce lands around Kathmandu seems rather high but then again it does provide a better road network around the city and one which may be of greater use and importance in the future.

Major Road Projects—Planned

Two major projects are being mentioned for the near future despite a declared decision to reduce investment in the Transport sector in the Fifth Five Year Plan as well as in subsequent plans. These projects involve the construction of a mid-Hill highway that would connect all four development regions. One section of it, possibly to be constructed by India, would connect Kathmandu (starting from Dolalghat) with Dhankuta on the eastern region. It is expected to be about 400 kilometers long and to cost over 100 million dollars. A preliminary survey of a possible alignment is said to have been completed but no further commitment has yet been made. Official statements suggested that the construction of the road could be completed within 10 years whereas other sources estimate that it would require 15 to 18 years. The second project, reportedly to be built by China, would link Pokhara with Surkhet. It would thus provide an extension further west to the already existing road from Kathmandu to Pokhara (also built by China). A preliminary survey of the possible alignment is said to have been completed and the road is expected to be 407 kilometers long costing approximately 100 million dollars. The estimated construction period is 10-12 years. The Chinese commitment seemed to be somewhat more sound that India's since an initial agreement for the construction of the road was signed between HMG and the PRC in February 1975.

The Hetaura-Kathmandu Ropeway

With the opening of the Tribhuvan Rajpath to traffic (1960), truck services replaced the old ropeway as the major means of transporting goods into Kathmandu. However, Kathmandu's growing economy and the increasingly larger volume of traffic on the Tribhuvan Rajpath soon required additional means of transport. A plan was proposed (1960) for the replacement of the old ropeway with a new bi-cable line. It was argued that with its potential capacity to produce electric power, Nepal should base some of its transport system on facilities operated by electric power, such as a ropeway.[47] It was believed that a new and technically more advanced ropeway would provide Kathmandu with less expensive and more efficient transport system. Moreover, calculations made with regard to expected rates on the ropeway suggested that its rates would be half those of the truck companies' rates. With these expectations the plan was approved (1960) and USAID undertook to construct the new ropeway from Hetaura to Kathmandu.

Completed in 1964 the new ropeway failed to fulfill its expected utility. The ropeway operation proved erratic from the very beginning and was impeded by both technical and administrative difficulties. The weakness of the basic supporting electrical energy system caused frequent power failures. The government's policies regarding the ropeway were vague and governmental support, when given, was inadequate. The ropeway company suffered from incompetent management as well as from severe competition by the powerful truck companies. Maintenance of the ropeway equipment was practically non-existent. Consequently, the ropeway company not only failed to obtain sufficient clientele even from governmental agencies, but also began losing the few clients that were using its service.[48] The company operated at only 30 percent of its total capacity, and the proposals that were made to remedy the situation were either ignored or deficiently implemented. The ropeway company failed to improve transport services in Nepal, but the recent increases in the price of oil could lead to some reconsideration of its potential utility.

Resources and Progress in the Transport Sector: 1951-1975

During the years 1951-1975 external assistance in the transport sector accounted for a large part of the total aid received by Nepal. Within the transport sector, road construction consumed close to 80 percent of the total investment. The availability of external aid to transport projects enabled Nepal to pursue a large scale development program and facilitated an increase of 850 percent in the total length of roads. The following discussion reviews the pattern of budgetary allocations in Nepal's development plans. It examines the sources of financial support as well as the progress attained in the construction of Nepal's road networks.

Resource Allocation

The First Plan, 1956-1960

In the First Plan, the transport sector was allocated Rs 104 million (or 31.5 percent of the estimated total outlay). Of this amount, Rs 50 million (48 percent) was originally allocated to road projects, Rs 33 million (32 percent) was set for railways, Rs 15 million (14 percent) for the Ropeway and Rs 6 million (6 percent) for aviation. In practice, actual expenditure during the plan period totalled Rs 214.5 million, and of this amount the transport sector's share was Rs 87.8 million (or 41 percent).[49] Within the transport sector itself actual expenditures were as follows—road projects consumed Rs 56.6 million (64 percent), ropeway Rs 25 million (28 percent), aviation Rs 5.1 million (6 percent) and the railway Rs 1.1 million (2 percent). Evidently, expenditures on road projects and

Table 1

**PLAN OUTLAY AND EXPENDITURE—DEVELOPMENT BUDGET—
PUBLIC SECTOR—TRANSPORTATION***
(in Rs Million)

<table>
<tr><th></th><th colspan="2">FIRST PLAN
1955/6-1959/60</th><th colspan="2">SECOND PLAN
1962/3-1964/5</th><th colspan="2">THIRD PLAN
1965/6-1969/70</th><th colspan="2">FOURTH PLAN
1970/1-1974/5</th></tr>
<tr><th></th><th>Estimated</th><th>Actual</th><th>Estimated</th><th>Actual</th><th>Estimated</th><th>Actual</th><th>Estimated</th><th>Actual</th></tr>
<tr><td>Total Plan Outlay</td><td>330.0</td><td>214.5</td><td>600.0</td><td>596.5</td><td>1740.0</td><td>1780.0</td><td>2570.0</td><td>3315.6</td></tr>
<tr><td>Allocated to Transp.</td><td>104.0</td><td>87.8</td><td>137.5</td><td>86.2</td><td>579.0</td><td>665.8</td><td>1010.0</td><td>1315.4</td></tr>
<tr><td>% Transp. of Total</td><td>31.5%</td><td>41%</td><td>23%</td><td>14%</td><td>33%</td><td>37%</td><td>39%</td><td>40%</td></tr>
<tr><td>Allocated to Roads</td><td>50.0</td><td>56.6</td><td>112.5</td><td>63.4</td><td>500.0</td><td>610.2</td><td>813.1</td><td>1002.8</td></tr>
<tr><td>% Roads of Transp.</td><td>48%</td><td>64%</td><td>82%</td><td>73%</td><td>86%</td><td>92%</td><td>80%</td><td>76%</td></tr>
<tr><td>Allocated to Railway</td><td>33.0</td><td>1.1</td><td>—</td><td>—</td><td rowspan="2">9.0</td><td rowspan="2">2.1</td><td rowspan="2">25.9**</td><td>—</td></tr>
<tr><td>Allocated to Ropeway</td><td>15.0</td><td>25.0</td><td>—</td><td>—</td><td>—</td></tr>
<tr><td>Allocated to Bridges</td><td></td><td></td><td></td><td></td><td></td><td></td><td></td><td>58.9</td></tr>
<tr><td>Allocated to Aviation</td><td>6.0</td><td>5.1</td><td>25.0</td><td>22.8</td><td>70.0</td><td>53.5</td><td>171.0</td><td>253.7</td></tr>
</table>

Sources: HMG, Five Year Plans I, II, III, IV; Ministry of Finance, Budget Speeches 1963-1976.

*The figures in this table refer to sums allocated to the transport sector per se; they exclude allocations made to communications and power, two fields which were also under the jurisdiction of the Ministry.

**Includes 23.3 to Nepal Engineering Institute and 2.6 to the Nepal Transport Corp. (NTC).

on the ropeway were higher than expected. The roads under construction during these years were--Hetaura-Narayangha; Kathmandu-Trisuli and Dhangari-Dandeldhura. It also included the completion of the Jogbani-Dharan Road and parts of the Tribhuvan Highway as well as construction of inner-city roads in the Kathmandu Valley.[50] Furthermore, from the "identity" of these projects it is also evident that external resources amounted to the larger part of this investment (see also Table 2).

The Second Plan, 1962-1965

This plan, prepared after the 1961 "royal coup," covered a period of three years (FY 1962/3 to FY 1964/5). The total outlay estimated for the entire plan was Rs 600 million of which the sector of Transport, Communication and Power, was allocated Rs 234.5 million (39 percent).[51] The transport sector per se was allocated Rs 137.5 million and included Rs 112.5 million (82 percent) set for road projects and Rs 25 million (18 percent) for aviation.

Once again, the plan was a broad list of projects and could not serve as a useful guideline for coordinated development effort. Accordingly the pattern of investment that actually materialized during this period had little to do with the original plan. In the transport sector it came to mean that the pattern of investment tilted once again, but this time in the opposite direction. The sector's share of total actual expenditure (Rs 596 million) declined to 14 percent (Rs 86.2 million) whereas the component "Power" consumed Rs 137.3 millions (23 percent). Nevertheless, of the total actual expenditure in the transport sector 73 percent (Rs 63.4 million) went into road construction and 27 percent (Rs 22.8 million) into aviation—e.g. more or less in accordance with the original allocation, percentage-wise.

This decline is explained by the fact that though verbal emphasis was put on the East-West Highway no "bidders" seemed to be interested. Furthermore, it was only toward the end of this period that several major projects were started and at that time they either consumed few resources or the actual expenditure involved in them were not included in the budget for that specific period. Those projects included the Arniko Highway (Kathmandu-Kodari) and the Raxaul-Bhainse portion.[52]

The Third Plan 1965-1970

The plan covering FY 1965/6 to FY 1969/70 had an estimated total outlay almost three times larger than in the Second Plan. It was based on more accurate statistical data and was prepared with a fifteen-year planning perspective. In this plan the original allocation to the transport sector was Rs 579 million (33 percent of the estimated total outlay) including Rs 500 million (86 percent) to road projects, Rs 70 million (12 percent) to aviation, and Rs 9 million (2 percent) to railways and the ropeway. It should be noted that estimated allocations within the transport sector were largely based on the Master Plan's recommendations. Furthermore, on the basis of past experience the Third Plan also indicated that some adjustments in cost estimates might be required as the plan progressed, particularly in projects involving construction.[53]

The pattern of actual expenditure that emerged by the end of the plan period was again somewhat different than originally projected. The total plan outlay increased from the original estimate of Rs 1740 million, and expenditures in the transport sector rose to Rs 665.8 million (37 percent). In fact, this was the first time that actual expenditure exceeded estimated allocation—both for the plan as a whole (by 1 percent) and in the transport sector in particular (by 15 percent).

In the transport sector, road projects expended Rs 610.2 million (91.7 percent), aviation Rs 53.5 million (8 percent), and railways and the ropeway Rs 2.1 million (0.3 percent). Review of the projects under construction during this period reveals that they included most of the major projects: the Sidhartha Highway (Pokhara-Sunauli); the Indian, Russian and British sectors of the East-West Highway; the Dhagari-Dandeldhura Road; the completion of the Arniko Highway, and the work done on the Kathmandu-Pokhara Highway.

The Fourth Plan 1970-1975

This plan differed from all other plans since it emphasized the completion of ongoing projects rather than the launching of new programs.[54] Resources were allocated accordingly, and of the estimated total outlay (Rs 2570 million) the transport sector was originally allocated Rs 1010 million (39 percent). In this sector, Rs 813.1 million (80 percent) was set for road projects; Rs 171 million (17 percent) for aviation, and Rs 23.3 million (2 percent) for Nepal's Engineering Institute.

The emphasis on completion of projects was particularly applicable to the transport sector since most projects in progress involved the establishment of links between numerous districts and zonal headquarters. As such, these projects were considered highly important from the administrative point of view and by the end of the plan period many of them were either completed or near completion.

In terms of actual expenditure the pattern that first emerged in the Third Plan became more apparent during the Fourth Plan. Total actual expenditure exceeded the original estimate by 29 percent (Rs 3315.6 million as compared with Rs 2570 million) for the whole plan, and by 30 percent (Rs 1315.4 million as compared with Rs 1010 million) for the transport sector. Of the total in this sector 76 percent (Rs 1002.8 million) was expended on road projects; 19 percent (Rs 253.7 million) on aviation; and 5 percent (Rs 58.9 million) on bridge-construction projects.

Budgetary Resources 1956-1975 and the Transport Sector (Table 2)

During the period 1956 to 1975 budgetary resources increased from Rs 330 million to 3315.6 million, an increase of over 1000 percent. These increases were made in practically every Development Plan. The largest increase was made in the Third Plan (1965-1970) with a total almost three times that of the Second Plan. Furthermore, while external resources amounted to more than three-quarters of the total during the 1956-1965 period, a significant decline in this proportion became apparent during the 1965-1975 period. Internal resource mobilization became increasingly more effective, to the point where in the Fourth Plan for the first time the volume of internal resources exceeded external resources.

This trend, however, was somewhat more limited in the transport sector. During the 1956-1975 period external resources amounted to over 80 percent of the total in this sector; most major projects were financed and executed by the donors' agencies and a few by multilateral agencies. Nepali organizations and agencies were mainly involved in the construction of local roads and several inner-city roads. Nevertheless, Nepali agencies gradually assumed a larger role in this sector and embarked on the construction of projects such as the reconstruction of the Hetaura-Narayangha road and the Nepalgunj-Surkhet road.

Progress in Road Construction 1951-1975 (Table 3)

In 1951 Nepal had one of the lowest ratios of kilometer of permanent road per 100,000 inhabitants (4.4) in the world. The total length of the roads in the country was 376 kilometers, of which only 88 kilometers were all-weather roads and 288 kilometers were seasonal fair-weather roads. Most of these roads were located in Kathmandu Valley and a few in the Terai. By 1956, prior to

Table 2

**INTERNAL AND EXTERNAL RESOURCES–
DEVELOPMENT BUDGET
(Actual, in Rs millions)**

	FIRST PLAN		SECOND PLAN		THIRD PLAN		FOURTH PLAN	
Total Plan Outlay	214.5	(100%)	596.5	(100%)	1780.0	(100%)	3315.6	(100%)
Revenues	22.0	(app.)	115.0	(19.3%)	741.5	(41.7%)	1817.0	(55%)
Foreign Aid			432.5	(72.5%)	1019.0	(57.2%)	1186.4	(36%)
	328.8*							
Foreign Loan			49.0	(8.2%)	19.5	(1.1%)	312.2	(9%)

Sources: HMG, Five Year Plans I, II, III, IV
Ministry of Finance, Budget Speeches 1963-1976

*Although the First Plan was originally projected to cost not more than Rs 300 million, the budgetary allocations during the plan's period totaled about Rs 600 million. Of this amount the actual expenditures were Rs 214.5 million. Evidently, the budget was drawn upon over-ambitious scale without considering the absorptive capacity. Additionally, since a large part of the foreign aid was earmarked to specific development projects, it is estimated that during the plan period, foreign resources amounted to 90 percent of the total actual expenditure. See HMG, The Three Year Plan, pp. 27-32.

Table 3

EXISTING ROADS IN NEPAL—PROGRESS MADE DURING 1951 - 1975
(In Km)

	1951	1952 - 1955			First Plan			1961 - 1962			Second Plan			Third Plan			Fourth Plan		
	Tot.	Tot.	%	Added	Tot.	%	Km Added	Tot.	%	Km Added	Tot.	%	Km Added	Tot.	%	Km Added	Tot.	%	Km Added
Roads	376	624	100	248	912	100	288	1090	100	178	1826	100	736	2730	100	904	3173	100	443
All-Weather	88 (24%)	259	42	171	343	38	84	420	39	77	436	24	16	1256	46	820			
Fair-Weather	288 (76%)	365	58	77	569	62	204	670	61	101	1390	76	720	1474	54	84			

Sources: HMG, Five Year Plans, I, II, III, IV

IBRD, A National Transport System for Nepal, 1965 (Vols. I, III)

The "Rising Nepal," May 21, 1976

the launching of the First Five Year Plan, the country reportedly had 624 kilometers of roads, of which 259 kilometers were classified as all-weather roads and 365 kilometers, fair-weather road. This increase is explained primarily by the construction of the Tribhuvan Rajpath, the Jogbani-Dharan Road, the work done on the Hetaura-Narayangha Road, and several inner-city and districts roads. This was an increase of 171 kilometer in all-weather roads and 77 kilometers in fair-weather roads. This increase in the all-weather road category as compared to the fair-weather road category would not be repeated in Nepal until the Third Plan.

In the First Plan and in accordance with the high priority assigned to the transport sector and road construction in particular, the target was set for construction of 1440 kilometers. Of this total, 480 kilometers were to be all-weather roads and 960 kilometers, fair-weather roads. By the end of the plan period the total length of roads was 912 kilometers of which 343 kilometers were all-weather roads and 563 kilometers fair-weather roads. A total of 288 kilometers was therefore added during the First Plan. It included 88 kilometers of all-weather roads, whereas the increase in fair-weather roads could be attributed to the work done on projects like the Kathmandu-Trisuli road, Dhangari-Dandeldhura (RTO), and the Janakpur area. During 1961-1962 the total length of roads increased to 1090 kilometers including the construction of 77 kilometers in the all-weather roads category and 101 kilometers in the fair-weather category.

The Second Plan called for the construction of 1440 kilometers of roads that would include 320 kilometers of all-weather roads and 1120 kilometers of fair-weather roads. The projects mentioned in this plan were: The East-West Highway, Kodari Highway (Arniko) and the Sunuali-Pokhara Highway (Sidhartha).[55] By the end of the Second Plan (1965) it was claimed that the total length of roads in the country was 1826 kilometers of which 436 kilometers were all-weather roads (i.e. an increase of 16 kilometers) and 1390, fair-weather roads (i.e. an increase of 720 kilometers).

This seems rather curious especially since it attributes the larger part of the increase (621 km) to work done on the East-West Highway.[56] Such a claim is surprising particularly in view of the fact that the Second Plan was only a three-year plan, during which actual expenditure on the transport sector was at its lowest. It should also be noted that actual work on the first portion of the highway (the far eastern portion) did not begin until 1966. Furthermore, even if we are to consider voluntary work done on this proposed highway, it remains doubtful that 621 kilometers were actually built in such a short period. These "mysterious" 621 kilometers were, most probably, century-old narrow foot trails and tracks that were "disguised" as fair-weather roads constructed during the Second Plan.[57] The claim that they were constructed during 1362-65 could only be made for internal political reasons since this highway was King Mahendra's "pet project" and had become a national symbol of Nepal's independence. Moreover, in its review of the estimated length of roads in Nepal in 1970, the Fourth Plan reported the total length of the East-West Highway to be 452 kilometers, of which 138 kilometers were fair-weather road and 314 kilometers, all-weather roads.[58] Progress achieved on this highway during the Third Plan, was stated to be 301 kilometers of which 121 kilometers were fair-weather and 180 kilometers were all-weather.[59]

Actual progress achieved in the Second Plan also included the work done on the Arniko Highway, several kilometers on the Sidhartha Highway, the work done in the Kailabas-Ghorani area and in the Jalkundi-Dang area as well as in the Kathmandu Valley.

The Third Plan projected the construction of 1365 kilometers of roads (758 all-weather, 607 kilometers fair-weather). By the end of the plan period the progress achieved amounted to 904 kilometers of which 820 kilometers were all-weather roads and 84 kilometers fair-weather roads. It was during this plan that most major road projects started. These included the Janakpur-Jhapa, Simra-Janakpur, and Narayangha-Butwal portions of the Mahendra Highway; the completion of the Arniko Highway; the work done on the Prithvi Highway, the Sidhartha Highway and the Dhangari-Dandeldhura road.

The Fourth Plan, which emphasized the completion of ongoing projects envisaged the construction of 1830 kilometers of roads including the completion of the western portion of the Mahendra Highway (520 kilometers). The progress achieved during the Fourth Plan was 443 kilometers, mostly all-weather roads. It included the Prithvi Highway (Naubise-Pokhara); the far eastern and the Dhalkewar-Pathlaiya (Simra-Janakpur) portions of the Mahendra Highway; it also included large parts of the Narayangha-Butwal portion and the Sidhartha Highway as well as parts of the Dhangari-Dandeldhura road. It did not include, however, any significant portion of the Western part of the Mahendra Highway.

During the last 25 years the total length of roads in Nepal increased from 376 kilometers to 3173 kilometers. About half of this total is comprised of fair-weather roads that are not motorable all year around. A relatively large number of districts and zones are now interconnected by roads, mainly in the Terai and the Central Development Region. The current network, although patterned in many ways after the Master Plan, differs from it mainly with respect to north-south roads in the Hills. These roads are yet to be constructed, though feasibility studies were already conducted on several of them. The extent to which these roads will actually be constructed depends largely on the availability of external aid, since HMG in its Fifth Five Year Plan (1975-1980) emphasized its intention to reduce investment in the transport sector.

REFERENCES

1. P.S.J.B. Rana and K.P. Malla (eds.); Nepal in Perspective CEDA, Kathmandu, 1973, p. 15.

2. International Bank for Reconstruction and Development (World Bank); A National Transport System for Nepal; Washington, D.C., June 1965, Vol. I, p. 22.

3. Ibid., p. 23.

4. CEDA; Study of the Transport Corporation of Nepal; CEDA, Tribhuvan University, Nepal, pp. 20-22.

5. IBRD; A National Transport System for Nepal; Vol. I, p. 24.

6. P.S.J.B. Rana and K.P. Malla (eds.); Nepal in Perspective; p. 16. See also L.E. Rose and B.L. Joshi, Democratic Innovations in Nepal: A Case-study of Political Acculturation, University of California Press, 1966, pp. 74-75.

7. His Majesty's Government, Nepal; A Draft Five Year Plan; Kathmandu, 1956, p. 39.

8. National Planning Council, HMG, Nepal; Three Year Plan (1962-1965), Kathmandu, 1962, p. 25.

9. Ministry of Economic Development, HMG, Nepal, The Third Plan (1965-1970), Kathmandu, 1965, pp. 21-23.

10. National Planning Commission, HMG, Nepal, The Fourth Five Year Plan (1970-1975), Kathmandu, 1970, p. 118.

11. IBRD, A National Transport System for Nepal, Vol. I, p. 1.

12. IBRD, A National Transport System for Nepal, Washington D.C., June 1965, 3 Vols.
Vol I, Report and Recommendations.
Vol. II, Supporting Technical Papers and Discussions.
Vol. III, Statistical Appendix.

13. Roads Department, Ministry of Transportation and Public Works, HMG, Nepal; A Revised 20-Year Roads Plan; Kathmandu, 1967.

14. United Nations Development Programme (UNDP); Road Feasibility Studies, Construction and Maintenance--Nepal: Project Findings and Recommendations, N.Y., 1975 (DP/UN/NEP-69-516/1), p. 1.

15. The Review is based on IBRD, op. cit., Vol I, pp. 31-41.

16. Ibid., p. 33.

17. Ibid., p. 35.

18. For further details on the road classification and the road authority, see IBRD, Ibid., pp. 35-39.

19. Ibid., p. 39.

20. Ibid., p. 40.

21. On the basis of transport planning and engineering considerations, the Mission divided Nepal into twelve geographical areas. In each of these areas the Mission recommended various roads and transport projects to be constructed. The details on each area and its suggested projects appear in IBRD, op. cit., Vol. II, Chapters 5 to 10.

22. It should be noted that several major projects started long before the plan was finalized. This placed some limits on the options open for the planners. Secondly, by submitting the plan it could not be expected that most projects would be actually implemented or, for that matter, that those projects recommended by the plan would be the ones implemented. The decision on which projects would be carried out depended on numerous factors other than developmental considerations. Consequently the shape of the secondary network would actually be governed by the primary network.

23. In this project the UN was designated as executing agency and the Ministry of Public Works, Transport and Communications as Government Counterpart Agency. The feasibility studies were carried out on contract by a consortium of consulting engineers--COALMA (Comtec. Alpine and Studio Macchi) from italy. In 1972 COALMA also received the contract to carry out the feasibility study on Kathmandu-Birganj Corridor.

24. This "corridor" known as the Tribhuvan Highway was, and still is the most important and heavily traveled primary road in Nepal. At the time the request was made the road became increasingly congested and difficult to maintain. For further details see "Major Transport Projects" in this chapter.

25. UNDP, Road Feasibility Studies . . . , 1975, p. 3.

26. The reports submitted by Coalma corresponded to each stage of the project; the reports were as follows:

(1) Nepal Road Feasibility Study, UN-HMG; Reconnaissance Survey, prepared by Comtec in collaboration with Alpine and Macchi, Rome and Kathmandu, December, 1970.

(2) Nepal Road Feasibility Study, UN-HMG; Part B; prepared by Comtec in collaboration with Alpine and Macchi, Rome and Kathmandu, May 1973.

(3) Nepal Road Feasibility Study, UN-HMG; Part C, Preliminary Investigation of the Kathmandu-Birganj Corridor, December 1971.

(4) UN-HMG, Road Feasibility Study for the Kathmandu-Birganj Corridor, March 1974.

27. The economic weight was evaluated as the average of five indices: the gross domestic product for 1970, the population in 1990, the value of agricultural production in 1990, the identified natural resources and the number of approved projects in the Fourth Development Plan. UNDP, Road Feasibility Study, op. cit., p. 9.

28. For details see Nepal Road Feasibility Study, UN-HMG, Part B, Projects 1 through 18.

29. UNDP, Road Feasibility Studies, Construction and Maintenance—Project Findings and Recommendations, N.Y., UN, 1975 (DP/UN/NEP-69/516/1).

30. Ibid., p. 12. For a detailed discussion see UN-HMG, Road Feasibility for the Kathmandu-Birganj Corridor, March 1974.

31. Ibid., p. 13.

32. L.E. Rose, Nepal: Strategy for Survival. University of California Press, Los Angeles and Berkeley, 1971, pp. 239-242.

33. For a detailed discussion on the economic impact of roads in that particular part of the country, see: Overseas Development Group (University of East Anglia), "The Effects of Roads in West-Central Nepal." A report to ESCOR, Ministry of Overseas Development. UK, 1977. See also M.C.W. Schroeder and D.G. Sisler, "The Impact of the Sunauli-Pokhara Highway on Regional Income and Agricultural Production." Nepal Occasional Paper, 32. Dept. of Agricultural Economics,Cornell University 1970.

34. H.C. Rieger and B. Bhadra, Comparative Evaluation of Road Construction Techniques in Nepal, CEDA, Kathmandu, 1973 (Preliminary Report, prepared for International Labor Organization, 4 parts), Part II, pp. 82-90.

35. E.R. Mihali, Foreign Aid and Politics in Nepal: A Case Study, Oxford University Press, London, 1965, p. 160.

36. Practically all the government's top officials interviewed, as well as the man in the street, emphasized time and again that this highway is a "source of pride" since it allows them to use "their" road rather than the Indian railways. This attitude toward the road prevails even though the highway is yet to be completed and regardless of the fact that it was built by "foreigners." However, Nepal's dependence on imported petrol from India and the cost of oil that would be required for transportation on this road could cause an even greater Nepali dependence upon India, thus making the Mahendra Highway a somewhat questionable source of growing Nepali independence.

37. R. Shaha, "Foreign Policy" in P.S.J.B. Rana and K.P. Malla (eds.), Nepal in Perspective, CEDA, Kathmandu, 1973, pp. 255-256.

38. H.C. Rieger and B. Bhadra, op. cit., pp. 102-104.

39. Asian Development Bank (ADB), Appraisal of Hetaura-Narayangha Road Project in Nepal, ADB, Report No. Nep: Ap-7, November 1972.

40. H.C. Reiger and B. Bhadra, op. cit., pp. 76-82.

41. See R.S.J.B. Rana, An Economic Study of the Area around the Alignment of the Dhangari-Dandeldhura Road, Nepal, CEDA, Kathmandu, 1971, pp. 1-28. See also USAID, "Engineering Evaluation, Western Hills Highway Project," prepared by Hoskins-Western-Sondregger, Inc., Lincoln, Nebraska, June, 1973.

42. The present alignment of the road came under frequent criticism with regard to its utility and potential economic contribution to the area. See Ratna S.J.B. Rana, An Economic study of the area around the alignment of the Dhangari-Dandeldhura Road. Nepal: CEDA, Tribhuvan University, Kathmandu 1971. For further discussion on the political and administrative aspects of the project see next chapters.

43. For example, see US AID, Ibid.

44. See HMG, Roads Department, A Revised 20-Year Roads Plan,

1967.

45. Nepal Road Feasibility Study, UN-HMG, Part B, op. cit., Project No. 17.

46. See US AID, op. cit., pp. 60-65.

47. CEDA, Study of the Transport Corporation of Nepal, CEDA, Kathmandu, 1973, p. 23.

48. Ibid., Chapter 1.

49. National Planning Council, HMG, Nepal, Three Year Plan (1962-1965), Kathmandu, 1962, p. 28.

50. National Planning Commission, HMG, Nepal, The Fourth Five Year Plan, Kathmandu, 1970, p. 123.

51. National Planning Council, Three Year Plan, p. 25.

52. Ministry of Economic Planning, HMG, Third Plan 1965-1970, Kathmandu, 1965, p. 6.

53. Ibid., p. 22.

54. National Planning Commission, HMG, Nepal, The Fourth Plan (1970-1975), Kathmandu, 1970.

55. The Three Year Plan (1962-1965), op. cit., p. 153.

56. The Third Plan (1965-1970), op. cit., p. 115.

57. No one among those interviewed including people who visited the area during the Second Plan period could recall the existence of 621 kilometers of fair-weather road along the proposed alignment of the highway or, anywhere near it, excluding foot trails and tracks around villages.

58. The Fourth Plan (1970-1975), op. cit., pp. 125-126.

59. Yet, in the estimated total length of existing roads, the Fourth Plan included the "missing" 621-kilometer roads. It was done "simply" by adding the reported progress achieved in each plan thus having the sum total of 3731 kilometers of road.

Planning Efforts During a Decade of Political Experimentation: 1951-1960

The 1951 "democratic revolution" was followed by a decade of experimentation with constitutional monarchy. It proved to be a decade of political instability affecting all branches of government and creating a paralyzing confusion within the civil service. In a period of less than ten years (1951-1960), twelve different governments were formed, including three occasions on which the monarch assumed direct rule.[3] In a country like Nepal which under the Rana regime experienced a century of relative stability, it was indeed an awakening to a totally new reality.[4] Political parties—a new phenomenon in Nepali politics—could not reach a consensus, no matter how ambiguous, upon the forms and rules they were theoretically committed to introduce. The administrative machinery, having a very limited capacity to perform large-scale tasks soon became further incapacitated by events in the political arena. Frequent changes in personnel, lack of skilled personnel, lack of administrative leadership, lack of consistent policies, and an increasing politicization made the bureaucracy anything but an effective apparatus to perform governmental duties. With not single institution capable of mobilizing sufficient support and resources to resolve the political power struggle or to contain its ill effects, the task of modernizing Nepal could not be performed in any satisfactory manner. Efforts made during the years 1951-1960 in the area of planning were no exception to this rule. Planning activities were of almost no consequence in determining actual development programs and planning units proved incapable of changing the situation.

In the Beginning there was to be an Organization

The call for modernization was made immediately after the overthrow of the Ranas. The new political leadership, united behind the necessity for changes, apparently decided that reorganization of the government machinery ought to be given first priority, and justifiably so. Having only limited capacity to carry out the task by itself, the Nepali Congress Government invited a team of Indian advisors to survey the system and submit its recommendations for administrative reform.[5] In early 1952 the team submitted its report which later became the basis for various reorganization schemes.[6] Among the recommendations that were immediately implemented was the establishment of a Ministry for Planning and Development.

The new ministry was expected to gather basic data on the economy as a preliminary step to the preparation of development plans. Yet, with the concept of "planning for development" meaning many different things to many different participants, such a task soon proved to be practically unimplementable. Lack of an adequate communications system and transportation network made the limited amount of available data highly inaccessible; cooperation among ministries was limited due to frequent organizational changes and the fact that there were not clear definitions of responsibilities and division of labor among governmental units. In additions, the Ministry for Planning suffered from lack of planners on its staff and often found it difficult to define its own tasks, let alone perform actual planning functions. The problem further worsened with the frequent changes in political leadership. The formation and dissolution of four different cabinets within a period of less than four years (1951-1955) produced greater confusion within the administrative machinery that had to cope with changes in personnel as well as lack of consistent policies. Under those circumstances it was not surprising that neither actual planning activity nor the establishment of a competent infrastructure for such activity in the future ever materialized.[7]

A New Approach, A New Agency

In 1955, following King Mahendra's ascent to the throne, a new and presumably more vigorous approach was taken toward the issue of development and planning. A grand scheme for reorganizing the government machinery was launched to correct various deficiencies evident in the functioning of the system.[8] A Royal Proclamation issued on October 9, 1955, emphasized the importance of proper planning and called for the preparation of a Five Year Plan. In fact, the success of a planning apparatus was recognized as dependent upon the proper execution of administrative reform. Yet, a constellation evolved that required the planning function to become "productive" long before the reform was actually carried out. Nepal, which joined the Colombo Plan in March 1952, was to participate in the Singapore meeting on October 17, 1955. The Ministry of Planning and Development was instructed to prepare a general draft outline of a periodic plan, in time to be incorporated in the Singapore deliberations. Within a matter of two weeks an outline was indeed prepared. The act, although a farcical exercise in planning, was nevertheless to become a new beginning. A year later, during the Acharya government's efforts to carry out the administrative reform program (assisted by teams of foreign advisors), a more substantial second draft of the periodic plan was prepared.[9] It became the country's First Five Year Plan. Owing to an acute lack of data, the plan did not serve as an outline for action in any sector of the economy,[10] but provided only a skeletal framework of broad objectives and priorities, emphasizing the urgent need to accumulate more data on various sectors and the importance of establishing supporting administrative units that would make future plans more detailed and pragmatic.[11] Other than that, it was merely a collections of public sector projects in different fields, with some estimates of costs.

In early 1957, with the administrative reform still underway, a partial step was taken to strengthen planning capabilities and improve coordination among development programs. A central planning unit, the National Planning Council, was established under the chairmanship of the Prime Minister. It became apparent that the Ministry of Planning and Development was too weak to perform its duties effectively and that the planning function could be improved only if an agency with solid political clout were put in charge. In the circumstances prevailing at the time, the office of the Prime Minister was among the more powerful. However, since not too many political actors wished to see the Prime Minister acquire additional power, the National Planning Council was established as a kind of "super unit" rather than an integral part of the Prime Minister's office. Its staff was comprised of members of different governmental and public institutions and its duties were vaguely defined: "to partake in the planning function, particularly in the area of policy formulation and coordination of programs."

It was not long before the newly formed agency demonstrated some basic weaknesses. With appointments to administrative positions in the Council often based on political considerations, a large proportion of the staff lacked the necessary skills to undertake planning operations. Furthermore, this style of recruitment enhanced fears among ministries that the Council was more than a mere administrative unit with the planning and coordination functions, and was politically motivated. The ministries refused to cooperate and resorted to tactics of delays, translating acts and proposals by the Council as maneuvers to either invade or control their own areas of jurisdiction. Having no actual controls over the flow of data from the ministries and lacking sufficient staff to process it adequately, the Council found itself incapable of performing its assigned duties. It could, of course, resort to the application of pressure,[12] but before such a step could even be seriously considered, a change took place and the issue became irrelevant.

In July 1957 K. I. Singh became the fifth Prime Minister since 1951, and one of his first acts in that capacity was the isolation of the National Planning Council from participation in planning activities. Singh, who considered the Five Year Plan to be a "farce" and viewed the NPC as "an agency established somewhat on the lines of a political circus," discarded the plan and used his cabinet as a planning agency for the formulation of a modified Two Year Plan.[13]

Published on September 1, 1957, the modified plan emphasized the development of transport and communication and proposed the establishment of some hydro-electric power stations and irrigation dams. No mention was made of the government resources available to support these undertakings. The plan, seemingly motivated by political considerations rather than by carefully thought-out economic considerations, never reached the stage of implementation since Singh's government was dismissed in November 1957.[14] King Mahendra, who assumed direct rule over the government, appointed a National Advisory Council to replace the cabinet and he also established a new National Planning Council under the chairmanship of his brother, Prince Himalaya (January 5, 1958). Singh's Two Year Plan was discarded and the newly formed NPC was assigned the responsibility of reviving the original Five Year Plan.

It was not long before the new NPC began to experience the same difficulties that had plagued its predecessor and to exhibit the same deficiencies and ineffectiveness for what seemed to be practically the same reasons. As with its predecessor, this NPC was also made a "super-unit" and was brought under the chairmanship of a person who enjoyed substantial political clout. Furthermore, this time it was made even less closely associated with the bureaucracy, having a member of the royal family in charge. The change, however, was more in form than in substance. The increasingly active role of the palace in the country's political process actually enhanced fears among the ministries that "politics was at the heart of the matter." The bureaucrats were reluctant to cooperate with the central planning unit; the Council continued to suffer from lack of skilled personnel; political appointees abounded; and there was no clear definition of its authority over governmental units. Difficulties in processing the inflow of data and the inability to effectively control the "production" of data undermined the Council's contribution to improved planning. Political clout, although a necessary condition for effective performance in the system, brought its own constraints and proved to be insufficient. When political interaction between the king and the parties intensified, the Council found itself virtually incapable of performing any planning function. Events in the political arena became the center of attention for all concerned, including the governmental machinery which was following the rule of "wait and see."

Intensified Political Activity Leaves Little Time for Planning Activity

On February 15, 1958, following demands from all parties, the king issued a Royal Proclamation setting February 18, 1959 as the date for the country's first general election. The Proclamation was accepted by all parties even though it stipulated that the election would be held for a Parliament rather than a Constituent Assembly and that the monarch would be the one to grant a constitution to the people. In May 1958, a new Council of Ministers composed of members of a four-party coalition was appointed, replacing the National Advisory Council.[15] With the establishment of this rather broad-based Cabinet, preparations for the elections intensified. Moreover, these preparations continued in an orderly manner despite the fact that one crucial factor remained unresolved—there had been no publication of the Constitution under which any elected government of the future would have to function. Indeed, it was only on February 12, 1959, one week before the elections, that the king proclaimed the new Constitution. The 1959 Constitution provided for a Parliament and a cabinet form of government, yet it clearly stated that the source of all legislative, executive and judicial authority was the Crown. The establishment of a democratic system was the most conspicuous omission in the statement of objectives.

Once the Constitution was announced elections began as scheduled on February 18, 1959. On May 27, 1959, following an overwhelming victory at the polls, B. P. Koirala and his Nepali Congress Party were called upon to establish a cabinet. The new cabinet was announced by the King and, when the new Constitution went into effect on June 30, the cabinet was transformed into a constitutional government.

With the establishment of an elected government and the inauguration of a representative legislature, Nepal entered a new era in its political development. The Nepali Congress government, commanding an absolute majority in Parliament, took a leading role in seeking to direct change in various areas of the Nepali system. It sought to introduce gradual but radical change in the system of land tenure and ownership; it initiated a series of attempts to mobilize the country's indigenous resources through the inauguration of the first income and property taxation program in the history of Nepal; it announced the appointment of District Development officers all over the country, giving them direct control over the administration of short-term development projects; and it moved toward overhauling the administrative machinery. In the area of development planning the government introduced an interim program to precede the launching of a Second Five Year Plan. This interim program was meant to be a preparatory step during which additional data on the economy would be gathered and an administrative reorganization scheme completed to establish "more favorable conditions" for the Second Plan. Formulated only in broad terms, the program never reached the stage of detailed specifications since events in the political arena brought an end to the elected government.[16]

On December 15, 1960, King Mahendra dismissed the Nepali Congress Government and introduced a direct rule under the emergency powers given to him in the 1959 Constitution.[17] The royal takeover, the fourth in less than a decade, was not immediately recognized as signalling the beginning of the end of parliamentary democracy in Nepal. However, subsequent events made it clear that this was indeed the beginning of a new political experiment, a new political system in which the monarch was to be the center of power.

Effective Planning as Dependent upon Consolidation of Political Power

Throughout the decade of political experimentation, planning remained a ceremonial, administrative exercise. In fact, during the First Five Year Plan, wide discrepancies between planning and actual operations were evident in practically every sector of the economy. The unplanned growth in the number of newly-opened schools and post offices far exceeded established norms, whereas the record in electric power development was abysmal. Construction of roads, inclusive of progress made in unplanned projects, reached approximately 30 percent of the planned target. A similar record emerged in the field of irrigation whereas in the milk and cottage industries production rose above planned targets.[18]

This limited progress in development programs, both in terms of actual accomplishments and the relevance (or lack of it) of planning to actual performance, could be partly attributed to the difficult circumstances under which Nepal found itself after the overthrow of the Ranas. A host of organizationally weak and ideologically confused groups, seeking to modernize a primitive subsistence economy and to democratize a traditional society were confronted by immense obstacles. There was a limited administrative machinery with some experience in government operations but comprised of the ousted Ranas and their followers. This administrative machinery, which for more than a century had been geared to attend to the Ranas' interests and function as the "family's" servants, proved to be totally unprepared to perform on a development-oriented, modern, large-scale basis. It also proved to be apprehensive about the democratic revolution and somewhat skeptical about its prospects of success.[19] The democratic forces, on the other hand, in addition to being relatively small in number, lacked skilled and experienced personnel to fully and effectively assume governmental administrative responsibilities. Furthermore, lack of basic data on the economy made formulation of realistic policies and the preparation of development plans most difficult. Lack of capable local administration in the various parts of the country, combined with the lack of a communications system and transports networks made the prospects of successful execution of plans rather slim.

The prospects for improved performance were further undermined by both the style in which the country's new political system developed and the forms of interaction that evolved between political parties and the bureaucracy. Organizationally weak parties and frequent conflicts within each party and between parties (often the result of personal rivalries rather than ideological differences) yielded growing uncertainty and confusion. No single party or a coalition of parties was capable of mobilizing sufficient support for its policies. Consequently, changes in cabinet membership became highly frequent. Second, the political parties concentrated their efforts to gain power in Kathmandu, adhering to the notion that "politics is the business of Kathmandu." Hence, efforts to organize grass-root movement were postponed "until such time as power is attained in the center," whereas competition over power in the center intensified. In this struggle for power, the bureaucracy became both the target and the tool. Appointments to government positions were not merely a way to reward party followers, but an essential practice employed by each party to secure its political survival as well as a power base. Thus, the size of the bureaucratic machinery increased rapidly and administrative positions were filled with unskilled officials.

The growing politicization of the bureaucracy and the ascriptive criteria for recruitment of personnel yielded a diminishing capacity to perform common administrative tasks. The frequent changes in the top echelons of the bureaucracy enhanced inconsistency in policies and entailed lack of administrative leadership. Issues related to job security, cooperation between ministries and the exchange of information between administrative units became overloaded with political overtones and power considerations. Skilled personnel were not necessarily assigned to units in their field of expertise, whereas units staffed with unskilled manpower were inclined to occupy themselves with the relatively safe practice of reshuffling papers. Schemes for administrative reforms, appropriate as they might have been, were usually defeated because they were perceived in terms of political power struggle. Moreover, the difficulties to establish a functioning bureaucracy enhanced the tendency to apply political criteria to administrative responsibilities and operations.

Under these circumstances, the consolidation of political power appeared to be among the essential preconditions to improved bureaucratic performance. Ironically, when the democratic process finally produced a force that seemed capable and willing to resolve this chaotic situation, the royal family found it incompatible with its interests. In 1959, following the first democratic elections ever held in the country, the Nepali Congress Party emerged with absolute majority in the Parliament. A cabinet comprised mostly of members of the victorious party was formed and began drafting its plans for changes in the socio-political system as well as for economic development. This powerful democratic force appeared far more threatening to the crown than a host of weak parties, especially in view of the fact that is became capable of placing the monarch on a platform above politics, into a position of reigning but not ruling. With the royal family's history under the Ranas and with an active monarch and skilled politician of King Mahendra's caliber, the crown apparently concluded that history would not be permitted to repeat itself within such a short period of time, at least not in as far as the position of the monarch in Nepal was concerned.

Planning for Development under the Panchayat System: 1961-1967

During the years 1960 to 1967 a new system of government was established.[20] The process was composed of two phases, the first of which began with the royal coup and ended with the promulgation of a new Constitution (December 16, 1962), and the second started from that date to the First Amendment to the Constitution on January 27, 1967. The first phase was characterized by an effort to eliminate elements of the previous structure (primarily the Nepali Congress Party though it later came to include a ban on all political parties) and a search for a new form of government. The second phase involved efforts to refine the fundamentals of the newly adopted system, "Panchayat Partyless Democracy", into an elaborate viable political-administrative structure.[21]

During this period the palace established control over the political process to the point where the system came to resemble an absolute monarchy. Moreover, with the ban on political parties effectively enforced, the vacuum created in the political arena was filled by the bureaucracy. Politics were rather rapidly incorporated within the administrative machinery and acts of political significance could be accomplished only through the palace or the bureaucracy. Administrative operations became the vehicle through which influence could be gained and power exercised. The monarch, in order to ensure that no other independent power base would emerge, used appointments to the administrative positions and a system of rotation among ministers for this purpose. A strong drive by the palace to control the bureaucracy was met by a display of bureaucratic responses—foot-dragging, cautiousness, unwillingness to assume responsibility for administrative acts, apathy among administrators, and a growing degree of corruption and inefficiency. Ministers and their own appointees realized that under the prevailing system their positions were inherently limited both in terms of opportunities to perform their assigned jobs and in terms of the time span at their disposal. A widespread practice emerged whereby many of the latter turned first and foremost to promoting their own particular interests while carefully avoiding any confrontation with the palace.[22] It meant, among other things, that the bureaucracy grew larger through extensive favoritism and parochial interests. It also meant a substantial increase in difficulties and built in constraints to bring about adequate cooperation among ministeries as well as an effective control by the palace of administrative operations. Furthermore, data provided by any given unit to either higher echelons within the same ministry or to other ministries was often inaccurate and at time fabricated; the same was even more true of information provided to the palace. This was partly due to the unavailability of accurate data and difficulties, administrative and technical, in establishing adequate communication systems. However, the practice widened and became a characteristic feature of the system, mainly because of growing politicization of the bureaucracy and the subsequent growth in corruption and inefficiency. The rule of thumb was—minimize the provision of data to the extent possible both because there were persuasive reasons to conceal accurate information on actual operations and acts and because such information could be used against you as political-administrative circumstances change, as they do almost constantly. Consequently, with accurate data being either nonexistent or withheld from decision-makers, decisions all too often proved to be detached from reality. Hence, on those occasions when decisions were followed to the letter, they rarely reduced the confusion within the bureaucracy; on many other occasions they were either partially disregarded or totally ignored, left to fade away in the bureaucratic maze.

The impact of these practices on planning efforts and the preparation of plans was substantial. It sustained the phenomenon of producing plans that were only remotely connected to reality and nurtured the gap between planning and implementation. Growing participation of external actors in Nepal's development effort introduced several new factors were into the system and affected some facets of the administrative process including planning. It involved, among other things, assistance in the accumulation of data and preparation of long-term plans in some sectors of the economy as well as provision of some pre-planned projects. Notwithstanding the positive effects of these contributions, they also carried certain adverse effects that further complicated the planning process in the country.[23] Still, during these years, 1960 to 1967, two consecutive periodic plans were inaugurated. These were the Second Periodic Plans, also known as the Three Year Plan in its totality (1962-1965) and the first half of the Third Five Year Plan. In both cases, planning remained inconsequential in determining actual development activity, and it was only in 1968 that certain changes were introduced to permit the beginning of a more promising trend.

Planning, Politics, and Administrative 1960 to 1967: New Structure, New Plans, and (Practically) the Same Old Constraints

On December 16, 1962, two years after the royal coup, a new constitution was promulgated providing the country with a new political system—the "Panchayat Partyless Democracy." During this interim period King Mahendra ruled armed with emergency powers and with a Council of Ministers of his choice. The direct royal rule emphasized two major themes—first, the elimination of elements of the previous system, primarily the Nepali Congress Party. Second, an intensive effort to secure a central role for the crown both during the interim period and in the proposed new system. Swift action was taken against the Nepali Congress Party on the day of the royal coup when the party's leaders were arrested. Three weeks later, in January 1961, a Royal Proclamation was issued declaring all political parties illegal. The swift and efficient manner in which the coup was carried out left the political parties leaderless and prevented the organization of effective opposition from any quarter.[24] Having paralyzed this group, the king then moved to eliminate the remnants of party support by initiating a major purge in the administrative machinery. Civil servants suspected of too close ties to the Nepali Congress Party were dismissed at all levels of administration and in all branches of government. The new appointments to higher administrative positions were openly based on political considerations, and merit in qualifications was at best an incidental factor.[25] Furthermore, the royal regime did not restrict itself to changes in personnel. It reformed the administrative structure superimposing in 1962 a new regional structure on the old system. This divided Nepal into 14 zones and 75 Development Districts which coincided with the territorial jurisdiction of the zonal and district Panchayats. A new Panchayat Ministry was given discretionary power over local (village and town) Panchayats, including the power to suspend or dissolve a Panchayat and to appoint a provisional Panchayat able to exercise full powers. And, as expressed all too clearly in the 1962 Constitution, the crown was established and identified as the only source of all legislative, executive, and judicial powers, the source of ultimate political authority.

On January 5, 1961, King Mahendra addressed the nation over Radio Nepal proclaiming a ban on all political parties. He charged the ousted government with "failure to achieve progress and development and creating a dangerous situation that threatened the future of the country." He then turned to list his charges, pointing out among others the deficiencies in planning. "There was no overall planning outlook, no regard shown to existing conditions, necessary resources, technical personnel and practicability, nor were schemes properly formulated. Individual and uncoordinated projects were taken up in haste. The result was that targets could not be achieved and our limited resources were wasted. Aid received from friendly countries could not be properly utilized as the ministers could not make the administrative machinery meet the requirements of the plan." In a concluding remark he added, "The present government (the Council of Ministers he appointed on December 26, 1960 under his own chairmanship) will bear these defects in mind and prepare a new plan that will be practical."[26]

The verbal emphasis on planning became one of the themes the monarch continued to stress during the immediate period after the royal coup. The fiscal year 1961-1962 was a non-plan year devoted to the introduction of organizational changes in the planning apparatus and the preparation of a better proposal for a new periodic plan. On February 16, 1961, the monarch dismissed the 1958 National Planning Council and established a new one under his own chairmanship. Linking the central planning unit with the "source of authority" was a symbolic act meant to emphasize the importance assigned to this function as well as an act intended to provide more teeth to the planning apparatus. Furthermore, the vice-chairmanship of the new Council was given to the Minister of Planning.[27] This composition could have been most appropriate for planning in Nepal since it combined both political clout (via the monarch) and professionalism (via the vice-chairman and his staff). Indeed, shortly after its formation the NPC, together with the Ministry of Planning, began preparing a new Three Year Plan which was launched in July 1962. The plan proposed programs for the fiscal year 1962 to fiscal year 1965 and was considerably more modest in scope than the preceding plans. It was "designed to make available basic data on the economy, to modify existing organizational structure and to lay down the infrastructure as will be required for sustained economic growth and in this manner to lay down a basis for more effective and comprehensive planning in the future."[28] In terms of long-range objectives, the plan emphasized the following: (a) the expansion of national production, (b) maintenance of economic stability, (c) expansion of employment opportunities, and (d) the establishment of social justice.[29] Priorities in the plan were "attuned to the main purpose and long-term objectives mentioned" and included "projects aimed at providing basic data on the economy, reforms in the organizational structure and building a basic infrastructure (transport, power and communication)."[30] Of significance to the planning function was the emphasis given to the need to reorganize the administrative machinery and, in that context, the high priority given to the introduction of more effective means for monitoring progress and detection of administrative bottlenecks. It included recommendations for the establishment of planning units within each ministry, the introduction of periodic progress reviews, and the assignment of a large coordinating role to the NPC.[31]

Sound as those objectives and recommendations seemed to be, many of them defied implementation. One observer commented that "The Second Plan cannot be evaluated in terms of targets because it did not have any,"[32] while one of the active participants in the preparation of the plan added, "at the time the Second Plan was prepared, knowledge of economic conditions in Nepal was still extremely limited . . . it was not possible, therefore, to set overall targets and goals in quantitative terms. The absence of data also prevented the formulation of a long-range perspective planning within which the Second Plan was to operate".[33] The lack of basic data on the economy was, however, only one of several obstacles in the execution of the plan.

The proposed reforms and the attempts to implement the decentralization policy (one of the fundamentals of the Panchayat system) created serious confusion within the administrative machinery.[34] Furthermore, numerous conflicts rose among administrative units over issues of jurisdiction and spheres of authority (which were only vaguely defined).[35] The political leadership underwent frequent changes both in terms of persons nominated to ministerial posts or relieved from office (June and September 1962, April and December 1962, February and July 1964, January 1965) and in terms of reallocation of portfolios (on all the former dates plus February 1961 and April 1964).[36] Under such circumstances the prospects for proper execution of an originally broadly defined plan were substantially reduced. The process of establishing a new political-administrative structure progressed in a manner which hampered effective coordination among various units of the bureaucracy and enhanced politicization of administrative acts and the growth of parochial interests. The establishment of Panchayat tiers required politicians to court their constituencies on a somewhat broader scale than in the past. Apparently it was translated into attempts to provide employment in the bureaucracy and to divert development programs into one's own constituency. Consequently, the bureaucratic machinery grew larger and competition over scarce resources grew fiercer undermining coordination efforts and planned development. Accordingly, one of the main feature of the Second Plan period was therefore the increasingly larger volume of governmental spending. In fact, development expenditure increased from Rs 70 million in 1962/63 to Rs 180 million in 1964/65.[37] Improved planning practices were not established and neither the recommendations to establish planning cells nor the call for the introduction of progress review were translated into practice.

Amidst the administrative reform and changes in the composition of the Council of Ministers, the NPC also underwent several changes. In early 1963 the monarch reorganized the NPC and appointed the chairman of the Council of Ministers, the presiding officer. The ministries for Planning and Economic Affairs were merged into one organization—the Ministry of Economic Planning. During this period the central planning organization began conducting several surveys on the economy (preliminary population census, sample agriculture survey, a national income estimate) in order to obtain a somewhat more accurate data basis for future planning. These activities which were preparatory steps toward planning, could not be accomplished in a satisfactory manner due to the aforementioned conflicts between governmental units and because of the lack of sufficient skilled personnel on the staff of the NPC. Furthermore, with only few exceptions, members of the NPC were themselves involved in the political-administrative power game. Some members created their own "individually protected spheres of influence," acting as patrons of certain constituencies and, on occasion, developing their own schemes and development plans with almost total disregard of coordination possibilities or developmental needs.[38] Being in a position where they were officially responsible for the preparation of plans and in substantial control of development expenditure, the members of the NPC and the agency itself became one of the more powerful units in the system. It thus led to increased interaction between the NPC and various governmental units as well as to increased development activity in some parts of the country. Yet, it was occasionally done with only limited regard for the planning function or in congruence to predetermined plans. Such a practice continued well into the early part of the Third Five Year Plan launched in 1965.

On the basis of data accumulated during the Second Plan period and the information gathered by the World Bank, the NPC, together with the Ministry of Economic Planning, prepared the third periodic plan. It was a much more ambitious and comprehensive plan, formulated within a framework of a fifteen-year perspective. A Royal Directive issued for the preparation of the plan stated that Nepal should attain within ten or twenty years the progress achieved by other nations as a result of centuries of industrial development.[39] Accordingly, the main targets of the Third Plan were fixed against a fifteen-year overall target of doubling the national income. The primary objective of the plan was "to develop the prerequisite for rapid economic growth." The social objectives were identical with those enumerated in the previous plan. In order to ensure rapid economic growth, the plan stressed the need to achieve certain specific objectives in some particular sectors of the economy. These included the introduction of land reform and institutional changes in agriculture, increases in foodgrain production, the development of basic industries (cottage and small-scale industries), and the construction of transport and electric power facilities.[40] The plan called for expenditure of Rs 2,500 million which would increase the total gross domestic product by 19 percent. Within this scope the plan included for the first time the local Panchayats and the private sector in addition to the public sector. The plan envisaged that some sixty percent of the expenditure would be covered by foreign assistance. However, a note of caution by S. B. Thapa, the chairman of the Council of Ministers and the NPC, was incorporated in the preface to the plan with regard to the formulation of the plan and prospects for its implementation. "It is not an easy task to formulate a nationwide plan with the limited information that we have on various sectors of the economy. It is, therefore, natural to come across shortcomings in the plan."[41]

With the advance of the Third Plan period, and particularly when it drew to a close in 1970, the "prophetic" remark on the plan's shortcomings proved true for various major objectives and sectoral targets. In assessing performance during the period of the plan, the Nepal Rastra (National) Bank reported: "There was only a slight shortfall in the financing targets of the plan but the physical targets lagged far behind."[42] Review of the progress in agricultural production indicated a ten percent increase in foodgrain production (as compared with the budgeted fifteen percent) while "the area utilizing improved seeds is only 13 percent of the targeted area, and the use of chemical fertilizers is about 17 percent of the target of the plan."[43] The record in irrigation showed that actual achievement was approximately 40 percent of the target and in electric power close to 35 percent.[44] Y. P. Pant, who was at one time an active participant in planning and the NPC, thus concluded: "Though the Third Plan was an improvement upon previous plans, on the basis of the experiences accumulated thus far it did not seem that development on the planned basis could grow very much beyond the initial stage of preparing the economy for the task of development. The base for more comprehensive plans could not have been established even so far, and our knowledge of the economy is still quite elementary."[45]

Three years after the Third Plan was launched it became apparent that targets proposed in the plan were overly ambitious and that implementation lagged embarrassingly behind the planned activity. A significant change was then introduced in the planning process in September 1968 as the NPC began revising the periodic plan and the revision entailed some downward adjustments of targets. This occurred, however, only after certain modifications were introduced in the political arena and subsequently in the role of the NPC.

On January 27, 1967 King Mahendra proclaimed the First Amendment to the Constitution which included about eighty amendments in various articles, some of which were gestures presumably indicating the "Liberalization" of the system, while others were means to further establish the monarch as the center of power. In the amended Preamble to the Constitution, emphasis was put on three aspects: (a) the partyless nature of the system, (b) the system's "democratic" nature, and (c) the principle of decentralization of administration.[46] In article 25, the functions of the Council of Ministers were extended from "providing the King with aid and advice" to exercising "general supervision and control over the administration of the country in accordance with the provisions of the Constitution, other existing Nepalese laws, and directives issued by His Majesty from time to time in the interest of Nepal and the Nepalese people."[47] It further provided that the King would appoint the Prime Minister and Deputy Prime Minister from among the Ministers of the Council--all appointed by him and responsible, collectively and individually, to the Crown.[48] The organization of nonpolitical associations was permitted.[49] The amended Constitution further provided for appointment of Zonal Commission in every zone in the country and vested in the king the right to make those appointments directly.[50]

By introducing the Amendment, King Mahendra seemed to be responding to unrest in the country while skillfully strengthening the power of the crown. Following the disastrous outcome of the June 1966 revaluation of the Nepali currency,[51] strong criticism had been voiced in the National Panchayat. King Mahendra, who sought to appease the critics, reorganized the Council of Ministers in order to include a recently released Nepali Congress Party detainee. Faithful to his practice of mixing various strategies, three days after the formation of the new Council, the King moved on another front and approximately 700 government employees were sacked, under the guise of administrative reform, without reference being made to the Public Service Commission.[52] Furthermore, a month later the king made yet another overture in the political arena by ordering the release of sixteen detainees, including two general secretaries of the outlawed Nepali Congress Party.

Additionally, he appointed, for the first time since 1960, a three-member Election Commission. Throughout November and December of 1966 several peasant and student demonstrations occurred; drought in the Terai further worsened economic conditions and the pressure for changes and relaxation of the system appeared to be growing. It was in this background that King Mahendra introduced the First Amendmnt to the Constitution and by so doing he apparently succeeded in neutralising opponents and critics, thus indicating the scope of his control over the system.

From Consolidation of Political Power to Consolidation of Administrative Control: Changes in the Planning Process, 1968-1975

Organizational and Procedural Changes, 1968-1975

The years 1968 to 1975 mark a period during which the planning apparatus was considerably enlarged both in terms of the number of participants involved in the function and the scope of operations, means, and procedures that it came to encompass. These changes, although still of limited impact on the relevance of plans to actual development activities, seemed to have enhanced a certain trend that might permit an effective incorporation of planning into the governmental process. The prospects of this development taking place depends on the manner in which it will be carried out administratively and, more importantly, on the extent to which those measures will be translated into terms of political power. The progress attained was accomplished without too many serious difficulties from the major political actors, though various political power considerations were involved in this process all along. Nevertheless, difficulties of this sort may arise at any point of time in the future, especially so if the palace would perceive the increased power of the Ministry of Finance a threat or an "unhealthy" trend in the political-administrative arena.

The changes that first led to this precarious pattern of development of the planning function began shortly after the promulgation of the First Amendment to the Constitution.[53] In May 1968, following the appointment of a new cabinet and due to the clearly unsatisfactory progress in the development effort, King Mahendra dismissed the National Planning Council and in its place established a National Planning Commission. The Prime Minister became the ex officio chairman of the Commission which also included five other members and a vice-chairman as its administrative head. A new detailed list of functions and responsibilities was prepared for the Commission in the areas of planning, supervision of implementation, and evaluation of progress in development programs.[54] Most importantly, the status of the NPC was specifically described as an advisory body.

In September 1968, in accordance with its new functions and following a Royal Directive, the new NPC began a process of revision of the Third Plan. Such a revision, which entailed some downward adjustments of the plan's targets was undoubtedly a significant change in the process of planning. It actually meant departing from the prevailing concept of "en bloc" planning (i.e. a plan taken as a whole for the total given period) and the beginning of a new pattern of "piecemeal" planning. Indeed, when the NPC began preparing the Fourth Plan, it further tried to establish this concept through efforts to institutionalize annual budgets as the basis for development spending.[55] This "innovation," which was intended to make planning more effective, was aimed at the reduction of the gap between plans and activity and was complemented by several additional measures. A call was made and directives issued to ministries and departments to establish planning units (a similar call had been made in the Second Plan, 1962, but was not implemented). These units were to be entrusted with the duty and responsibility of preparing the plans, annual programs, and budgets of their respective organizations as well as to evaluate progress in implementation of their programs. Plans, proposals, and reports were then to be submitted to the NPC which would coordinate and consolidate them into an overall plan. The NPC was also expected to make regular evaluations of the progress of the plan's programs and to obtain semi-annual progress reports from all governmental agencies and units. The NPC would then review the reports and submit its findings and recommendations to the Cabinet for approval. Furthermore, in order to "improve coordination among ministries and provide for more effective utilization of external resources," the Foreign Aid Division was transferred from the NPC to the Ministry of Finance. This organizational change thus meant that direct control over external resources would be exercised by one ministry alone (as compared with the practice under which different ministries and the NPC were dealing with various aid-donor agencies) even though the latter were expected to "consult with and seek the advice of NPC." Other changes were proposed in budgeting procedures[56] and a new emphasis was put on regional development approach and the concept of "Growth Centers."[57]

The Fourth Plan (1970-1975)

In view of the record of the Third Plan, the NPC prepared the Fourth Plan with specific emphasis on the completion of projects started in earlier plans.[58] Incorporating the various ideas and innovations generated toward the end of the previous plan, the main objectives for the Fourth Plan were stated as follows: to maximize output; to establish the base for sustained and long-term economic growth (transport, communication, electric power); to expand and diversify international trade; to secure development and stability by controlling price levels; to make effective use of manpower resources and to control population growth; and to create conditions conducive to the emergence of a society free from exploitation.[59] For the attainment of those objectives the plan proposed the following policies: mobilization of internal resources (increasing the level of savings in the private sector and increase in investment rate); concentration of limited resources (creation of Growth Centers in selected regions according to their potential for development and in order to achieve proper utilization of scarce resources and create a balance between regional and national development); incentives for investment in the private sector (provision of infrastructure and government-established industrial projects); diversification of foreign trade (export goods and agro-based industries); and social justice (changes in the tax and revenue system, strengthening small industries and cooperatives).[60] The sectors emphasized in the Fourth Plan were similar to those in previous plans, namely, transport, communication, and electric power (seen as an essential infrastructure for accelerated growth in both agriculture and industry), agriculture development and industrial growth.[61] The plan envisaged a total outlay of Rs 3,450 million (as compared with Rs 2,500 million in the Third Plan), of which 2,540 million were to be invested in the public sector.

The emphasis on the down-to-earth approach that presumably characterized the plan ran aground shortly after it was launched, shedding further light on the constraints on effective planning in the country. Trade and transit negotiations with India and certain "pressure" methods employed by the latter created a scarcity of building material and fuel, thus causing serious delays in various projects. Poor harvests for three successive years further reduced prospects for attaining the set targets. Increased participation of aid-donor agencies in Nepal's development effort enlarged the scope of confusion among Nepali administrators to the point where one observer commented: "How can you have long-range planning when a country depends on approximately twenty foreign donors, each of whom has its own priorities and, what may be worse, none of which is likely to be identical with those of Nepal?"[62] It was also pointed out that "failure to prepare the cost estimates, survey reports, designs, acquisition of lands and timely recruitment of staff have been the recurring phenomena."[63] And, as with all previous plans, difficulties arose frequently in the attainment of proper coordination among different ministries. Furthermore, during the plan period, starting from July 1970 there were six occasions when the composition of the cabinet changed. Prospects for proper implementation of the plan under those circumstances were at best low. Indeed, in its mid-term appraisal of the progress achieved during the first two and a half years of the plan period, the NPC provided a long list of unaccomplished targets. It included numerous projects in all the sectors that were given high priority in the plan (agriculture, land reform, transport, communication, industry and commerce, and power.)[64] The appraisal pointed out, in general terms, the weakness of the administrative machinery, deficiencies in project planning, insufficient and ineffective supervision, and lack of adequate coordination among ministries. It did not (and for many "practical" reasons could not) appraise the adequacy of the original targets and objectives that were at the very basis of the plan.

The formulation of the Fourth Plan, although more elaborate than previous exercises and based on a relatively larger input from the "line" ministries, remained in essence a process of planning from the top down. It actually sustained the phenomenon in which implementing agencies felt little commitment to the plan's goals, not having a share in the decision-making process. The innovations proposed at the start of the plan period were difficult to implement since change toward annual budgeting seemed to be, in the eyes of the ministries, an effort by the central authorities to increase their control over the formers' operations.[65] Moreover, annual programming presupposes capable administrative machinery endowed with adequate coordination and flow of information among units. In Nepal, these conditions simply did not exist and the proposed innovations were slow to generate it. Second, for the purpose of effectively introducing the planning/programming innovations there had to be both sufficient skilled personnel in each ministry or department and a stable central planning agency capable of assisting the former as well as making adequate use of data input received. The manner in which planning cells evolved in most ministries suggested compliance with the requirement to establish such units and, at the same time, isolation of those units from participation in the organization's actual activities. Planning cells were usually staffed with insufficiently skilled personnel and were often cut off the main line of communication, without having any means of control or verification of data they were receiving. In that respect, those units were considered an unavoidable nuisance not to be over-concerned with.

The NPC, on the other hand, stripped of many of its powers, faced with an uphill contest with the Ministry of Finance, and having the experience of previous plans, understood all too well that a "plan" could not be a comprehensive program for action. Irrespective of the adequacy of the "product" (i.e. the plan), it could be expected that substantial changes would be introduced in it during the process of implementation. Some of those changes might be caused by unforeseen factors such as the negotiations with India, flow of foreign aid, or weather conditions; most other changes could be expected to evolve from the familiar factors of limited administrative capacity and the struggles between the bureaucracy and the palace and among units of the bureaucracy itself. Consequently the NPC chose to provide a detailed list of policies and targets in sectors or program where it had a better data basis. In other sectors and in the area of general objectives it simply sought to satisfy the "elegance of literary presentation", regardless of actual prospects for its implementation.

Different participants and observers pointed out this phenomenon and its ramifications. R. C. Malhotra wrote, "development planning in the past has been somewhat like a ceremonial exercise, performed with great gusto and fanfare every five years, but then relegated to the background . . . "[66] Pashupti Shumshere J. B. Rana, in his criticism of the Fourth Plan, examined each of its major objectives and concluded[67] that with regard to the emphasis on completion of projects, "There are several projects being continued or completed in the Fourth Plan that have never been visualized in any earlier plan . . . . Should we not admit the fact that many of the projects being completed are those which have been decided on an ad hoc basis, without the disciplined decision-making of national planning and without the consideration of the cost and benefit of possible alternative courses of action." With regard to the prime objective of maximizing output, he said, "I cannot understand why establishing the precondition for sustained economic growth should not have been ranked as the primary objective . . . when that is the way investment has been allocated." With regard to the four per cent per annum rate of growth he wrote "Long gestation period investment is by definition investment that will not raise productivity in the plan period . . . (thus) if we are to achieve a 4% p. a. presumably we must achieve it with the quick yielding investment which is 123.46 crores . . . this outlay in the plan was expected to produce an output higher than the total investment (in fact 131.2 crores)." With regard to keeping the price level, "due to the open border and the fact that the vast majority of our trade is with India, both the prices of our exports and imports are largely dependent for determination on the Indian price system . . . I could not find (in the plan) a correlation between estimated price rises in India and our own rate of growth." And finally, with regard to emphasis on labor intensive methods, "the choice of techniques in transport projects (the largest of all sectors) is determined not by the Nepalese government but by aiding organizations."

This criticism, although made only one year after the launching of the Fourth Plan, remained valid throughout the plan period. The innovative propositions made prior to the formulation of the plan could not be effectively translated into practical measures to improve performances during the plan period. Some of those became operational in time for the preparation of the Fifth Five Year Plan (1976-1980). It should also be emphaszied that the manner in which several of these changes evolved could just as well undermine the rationale behind the innovations and their potential effectiveness. The issue remains mostly a political-administrative one and rarely becomes a purely technical-administrative matter. In order to examine these developments let us turn first to a discussion of the current planning procedures and the major participants in the planning function.

Major Participants and Planning Procedures, 1968-1975/76

Responsibility for planning in Nepal is under the jurisdiction of some of the highest-level bodies in the country. It involves the palace and its special units; it is carried out by national advisory councils; and it is part of the responsibilities of practically every ministry within the bureaucracy. The division of labor regarding this function seems to be arranged along hierarchical lines, starting from the king and the advisory bodies established by him down to planning cells within ministries and departments.

The National Development Council

Established in June 1972 under the chairmanship of the King, the National Development Council is the highest forum for the formulation of development policies and periodic plans.[68] The Council is composed of all the Ministers of the Cabinet, the Chairman of the Standing Committee of the Council of State (Raj Sabha), three to five former Prime Ministers or former Chairman of the Council of Ministers (all nominated by the King), Chairmen of the standing committees of the Rastriya Panchayat (National Assembly), all members of the National Planning Commission, National Chairmen of the Class and Professional Organizations, Chairmen of the fourteen Districts Panchayats (one from each zone, serving on the council for one year), President of the Nepal Federation of Industries and Commerce, and three to five distinguished persons (nominated by the King).[69]

The Council, whose functions are decided by the King, is essentially a national advisory body to the King on matters of development. The Council reviews national objectives and issues guiding principles to the National Planning Commission (NPC) regarding the preparation of periodic (three to five years) plans and annual development programs. It conducts annual evaluation sessions to review progress made in development programs and to issue directives on steps to be taken for proper implementation of those programs. Additionally, if the monarch should so command, the Council may conduct inquiries into "laxity, negligence, corruption, and other evils which hamper the proper implementation of plans . . . (and) submits reports to the King along with recommendations to punish the guilty persons or officers, if necessary, and/or to confer rewards, if so appropriate." The Council meets at least once a year, normally in the middle of the fiscal year, with a view to be able to issue necessary directives to the NPC for the next fiscal year.

The National Planning Commission

The body officially responsible for the preparation of periodic plans and annual development programs is the National Planning Commission (NPC).[70] Headed by the Prime Minister as its ex-officio Chairman, the NPC is an advisory body not granted decision-making power. It has a full time Vice Chairman as its administrative head, three full-time members, one member-secretary, and two ex-officio members—the Chief Secretary and the Secretary of the Ministry of Finance respectively. The Vice Chairman and the members are appointed directly by the King. The Prime Minister is appointed to his position by the King and ex-officio serves as the Chairman of the NPC. The Chief Secretary is similarly appointed by the king and serves on the NPC ex-officio.

Subject to the guidance and directives of the NDC, the functions of the NPC are as follows:

A) <u>In the area of plan formulation</u>

1) to make recommendations to HMG on policies relating to economic and social development.

2) to formulate long term objectives of national development, and on their basis to prepare periodic plans, specifying priorities of development.

3) to formulate annual plans within the framework of the periodic plans and to recommend to HMG on their implementation by making budgetary projections.

4) to issue directives to concerned Ministries, Departments, and Public Sector Agencies regarding preparation of the various plans referred to in (3).

5) to provide technical assistance on plan formulation to the Planning Cells of different Ministries.

6) to collect necessary statistical data and to carry out required research for the formulation of national plans.

B) In the area of resources planning

1) to estimate annually internal and external resources and to suggest ways and means to increase them.

2) to assess manpower requirements of plans.

3) to estimate necessary development and construction materials for projects incorporated in the plan.

C) In the area of plan implementation and evaluation

1) to continuously supervise implementation of development projects for compliance with approved program and budget; and to assist in overcoming shortfalls and difficulties encountered during implementation.

2) to make semi-annual evaluations on the progress in implementation of different development projects and to submit to the Council of Ministers a report along with recommendations for greater efficiency or changes in programs and policy.

3) to assist the National Development Council on formulation of regional policies by making studies on potentialities of various regions of the country.

4) to organize and conduct Seminars and Conferences periodically for the officers working in the Planning Cells of various Ministries, Departments, and Government Agencies.

5) to publish data and information on objectives, programs, and achievements of national plans, so as to keep participants and the public at large informed of trends and targets.

The NPC is comprised of four Divisions, each headed by a Joint-Secretary. The Divisions are (1) Sectoral Program Coordination (2) Plan Implementation and Administration (3) Resource Planning and (4) Evaluation, Economic Research and Regional Policy. Additionally the NPC also includes the Central Bureau of Statistics.

Ministry of Finance

Probably the most powerful among all Ministries, the Ministry of Finance was responsible for the formulation of national fiscal policies.[71] It prepared both the regular budget and the development budget and exercised control over "foreign aid" resources. In that respect the Ministry of Finance assumed the role once held by the NPC and was actually responsible for the preparation of developmen programs. The general guidelines regarding priorities and estimated budget allocations for the forthcoming fiscal year were prepared by the NPC and the Ministry respectively, but within this general framework the Ministry apparently enjoyed substantial leverage that actually allowed it to determine the scope and nature of most annual development programs. In addition to the Foreign Aid Division and the Budgeting and General Accounting Department, the ministry is composed of the Departments of Land Revenue, Taxation, Customs, Excise, Mint, and Research. The Programming Section of the NPC was officially transferred to the Ministry of Finance, providing the ministry with the broadest jurisdiction over both budgeting and programming.[72]

The Ministries and quasi-governmental organizations prepare a detailed outline of their respective development plans, projects, and programs. Planning cells were established in most of the ministries during 1970-1975, whereas planning units within the departments (ministerial sub-units) still await introduction into the system. The function of the ministerial planning cells are as follows: (1) preparation of plans and consolidating of studies and plans prepared by the departmental planning units; (2) assistance, supervision, and coordination with regard to work performed by departmental planning units; (3) evaluation of ongoing projects and preparation of the semi-annual progress reports which are submitted by the ministeries to the NPC. Officially all work related to planning and evaluation is to be done under the direction of the NPC and with its technical assistance. Thus far effective functioning of the planning cells has proved to be most limited.[73]

Units of the Royal Palace

In addition to the planning input made by the aforementioned institutions, various planning functions are also performed by the Royal Palace Services.[74] The Palace units, primarily the Royal Palace Secretariats and the Janch Bujh Kendra (JBK—Research and Enquiry Center) are "staff" agencies directly responsible to the King. The principle function of the Royal Secretariats is to act as liaison with the various ministries, with a view to follow up on Royal Directives issued to those ministries.[75] As such and due to their "organizational" location at the center of decision-making, the Secretariats play a crucial role in the decision-making process. They are the monarch's close mentors and advisors as well as his own independent coordination and supervising unit. A similar though somewhat more clearly defined role is being performed by the JBK.

The Janch Bujh Kendra, originally conceived as an "Ombudsman" office, was established in January 1971 by the late King Mahendra.[76] The JBK was put under the supervision of Crown Prince (now King) Birendra. At first, it acted primarily as a special investigation unit, looking into public complaints against senior public officials. However, when King Birendra ascended the throne the JBK's investigative responsibilities were enlarged. It was actually made both an Ombudsman office and the Palace's special investigation unit entrusted with responsibilities to look into all sorts of maladministrative behaviour. As such, it could conduct inquiries on its own (i.e. the palace) initiative in addition to those conducted in response to public complaints. Furthermore, the JBK was also assigned "research responsibilities to conduct studies on various aspects of national development policy and administration, with a view to make recommendations for changes in policies and administrative reforms."

External Participants

Policy formulation and the planning process in Nepal also involved external participants to a degree sometimes uncommonly high even for Third World countries. The country's dependence on foreign aid, combined with the lack of technical capability and skilled personnel in several areas, often creates a situation in which the aid donor, either intentionally or by necessity, finds itself playing a major role in the formulation of development policy. Such a situation sometimes evolve due to the donor's interest in "selling" its product (e.g. China and the Kodari-Kathmandu Road) or due to a certain Nepali attitude which because of indifference allows the donor to play this role, filling the vacuum in a certain area of policy-making. In some programs (like health services or dairy industry) HMG realizes that it either cannot or does not wish to be fully involved in complex planning exercises. In these cases, fully planned programs backed by adequate financial support may be accepted even if they are not congruent with the established order of priorities for development.

Planning Procedures

Current planning procedures in Nepal are based on two major principles: first, the basic framework for development planning in the country is a periodic plan, usually a Five Year Plan; secondly, annual plans serve as a guideline in the implementation of the periodic plan and as a means to secure proper adjustment in the plan according to changing circumstances. The introduction of annual plans began during the Fourth Plan period and became a planning principle during the preparation of the Fifth Five Year Plan, 1976-1980.

In comparison to previous plans, the preparation of the fifth periodic plan stands out as an elaborate process involving a large number of participants as well as newly adopted planning principles. The manner in which it was done is as follows. In February 1973 the newly created National Development Council was convened under the chairmanship of the King for its first session, to determine the basic principles of the Fifth Plan. Directives were then issued to the National Planning Commission regarding national objectives and priorities for the 1975-1980 period. The NPC in turn provided the "line" ministries with guidelines for the preparation of their respective periodic plans. In September 1973 the NDC met for its second session in which it reviewed progress made during the fiscal year 1972-1973 and was also presented with a report by NPC on the national economic situation. In November 1973 the NPC began its deliberations on working papers prepared by the ministries regarding their proposed plans for the Fifth Plan. In June 1974, following additional drafts prepared by the ministries, the NPC began its deliberations on the financial resources for the Fifth Plan. A draft of the total financial outlay for that plan was published by the NPC in October 1974 in conjunction with the NDC's third session. It was followed by publication of working papers prepared by the NPC in several fields—housing and physical planning, trade, irrigation, civil aviation, telecommunication—for the Fifth Plan. Additionally, deliberations on financial resources continued between the NPC and the ministries. By June 1975, shortly before the beginning of the Fifth Plan's period, the NPC approved the final draft for that plan. The final draft was then presented to the fourth session of the NDC convened in July 1975 (i.e. the beginning of the plan's period).

The steps taken in the process of preparing the Fifth Periodic Plan, the extent of input officially sought from the ministries, and the length of time allocated for the preparation of the plan, all stand out as substantial improvements to previous planning exercises in Nepal. Yet it seems as though the more significant change in planning procedures was in the incorporation of annual plans into the process. This change took place twice, once during the early stages of the Fourth Plan and again shortly before the launching of the Fifth Plan, reflecting changes in planning responsibilities of the NPC and the Ministry of Finance.

Ever since the introduction of annual budgets as the basis for development spending, the NPC was entrusted with the responsibility of preparing development programs. The process of preparing the plan was as follows. On the basis of the semi-annual reports and according to the objectives stated in the plan, the NPC prepared guidelines and a set of priorities for the upcoming fiscal year. Subsequently and in cooperation with the Ministry of Finance, a preliminary budget proposal was prepared. In December general guidelines regarding priorities and estimated budget allocations were issued to the ministries together with forms for proposed programs. The ministries, on the basis of those guidelines and with proposals generated within their own organization, prepared their respective annual prgrams including budget estimates for each project. Once completed, usually during the period from January to March, the proposed plans were sent back to the NPC for review. Items and budgets in the proposals were then discussed in a series of meetings between the NPC, representatives of the concerned ministry, and representatives of the Ministry of Finance. Following this, the NPC was to finalize the annual program for each ministry and submit it to the cabinet for review. The cabinet, expected to approve the plan in principle, could nevertheless make changes in the budgetary allocations for the programs included in the plan|

Recent organizational and procedural changes regarding the preparation of annual plans altered the aforementioned process. In the middle of the 1975-1976 fiscal year the programming division of the NPC was transferred to the Ministry of Finance. Consequently, responsibility for preparing annual plans are now being shared by the two, with the latter given the more substantial role. Similar to the previous practice, the NPC sets priorities and prepares guidelines for the annual plan. At the same time, the Ministry of Finance prepares the annual development budget which predetermines allocations to each ministry and occasionally encompasses earmarked allocations to specific programs within the ministries. In December the Ministry of Finance issues forms to the various ministries for the preparation of their annual programs. The ministries are expected to follow the NPC's guidelines and prepare their plans according to the budgetary allocations determined by the Ministry of Finance. Once completed, during the period from January to March, the forms are sent to the Ministry of Finance for review. The proposed programs are then discussed in a series of meetings between the Ministry of Finance, representatives of the concerned ministry, and representatives of the NPC. By June the Ministry of Finance finalizes the plans and submits it to the cabinet and the Rastriya Panchayat for approval in time for the new fiscal year (July).

The larger share of responsibility for the preparation of annual development programs now falls under the jurisdiction of the Ministry of Finance, the role of the central planning agency having been reduced.

How Much Difference Makes a Difference—An Improved Planning Apparatus?

Formal organizational changes are, most frequently, either means employed toward the attainment of a certain goal (or goals) or the end-result of such a process. In Nepal, the "innovations" introduced in the planning function were only partly aimed at the attainment of improved planning and remain far from being major determinants of the prospects for improved planning. The measures adopted emanated from the political-administrative arena and were largely determined by political-administrative considerations. In that respect, these measures were used as a means to attain various goals, one of which was improved planning. Similarly, the effectiveness of these means in the planning process will be determined by the impact they may have on the political-administrative arena and subsequently by essentially the same or similar considerations. The issue involves the inter-relationships between the palace and the bureaucracy and within the bureaucracy itself among the various units of which it is comprised. Additionally, though to a somewhat smaller extent, it involves aid-donor agencies and their modes of operation.

The changes introduced in the planning apparatus during the years 1968-1975 provided for the establishment of certain necessary conditions for improved planning. However, they also set some of the major participants on a potential collision course that can cause a serious setback to planning efforts in the country. In particular, this involved a growing participation of the palace agencies in detailed planning operations, and the placing of substantial planning responsibilities and powers in the hands of the Ministry of Finance. The relationships between these two and the manner in which they develop will determine the outcome of the innovations adopted and the role other actors may come to play in the planning function.

A Growing Involvement of the Palace

The establishment of the JBK seems to be another move by King Mahendra to improve his control over the political-administrative system. He apparently realized that the development effort was lagging too far behind stated objectives and that data supplied by the bureaucracy to the palace on various development programs was grossly inadequate and inaccurate. Knowing all too well the causes for the ministries' reluctance to cooperate in the matter and the ineffectiveness of the prevalent control mechanisms, he proclaimed the establishment of the JBK as a new ombudsman-like office in the palace. What it eventually amounted to was the creation of a parallel planning and evaluation unit in the palace. The JBK, armed with the legal authority to ask for files and information from any government unit and staffed with well-educated young officials, soon became a synonymous name for trouble in the eyes of many bureaucrats. In addition to conducting investigations into improper administrative practices, the JBK—by virtue of its staff and mode of operation—proved capable of processing information reaching the palace into more or less coherent files on various policy areas and sectors of the economy. Consequently, it provided the palace with a data basis that enabled the king to ask somewhat more penetrating questions in some areas of governmental activity and to monitor progress in matters he considered important. However, such a means could not be used on a very large scale. Having a relatively small staff, the JBK obviously could not control the vast domain of the bureaucracy. Additionally, an over-utilization of the JBK could increase the bureaucracy's animosity toward the palace and further reduce the cooperation between the two.

Indeed, in recent years and particularly since 1974, the JBK became increasingly involved in preparation of policy papers and proposals for programs, acting mainly as the palace's special think-tank. In that respect a problem-solving attitude by the palace somewhat increased prospects for improved planning in the country. This, however, is not to suggest that the issue of politics in administration diminished significantly or that the above-mentioned trend is sufficient to set in motion a similar trend in the bureaucracy. In fact, ministries complained of not being consulted on plans and programs officially under their jurisdiction. The criticism is often directed against the JBK who, according to the critics, "does it all and leaves us with orders to implement those plans."[77] Moreover, other developments in the political-administrative arena seem to sustain the situation in which contrasting trends emerge simultaneously.

The establishment of the NDC is another example of this case. As a new forum for the formulation of development policies, the NDC seems to be a farce. It is first and foremost a political body controlled by the palace and it lacks the ability to be anything but a debating forum or a rubber stamp for decisions made elsewhere. It is too large to be effective even though or perhaps because of the fact that it includes almost all the major political figures in the country. Moreover, the NDC appears to be a means by which political figures who do not have any official position at a given time are kept in line in this "terminal". For those who have already held high-level government positions, the NDC is a "transit station" in which they are reminded where their appointments come from and what is required from them for future appointments. In that respect positions in the NDC keep the politically active elite away from other political activities that might be unfavorable to the monarchy. For those who are on their way up the ladder, appointment to the NDC is a promotion and a springboard to other high-level positions. Finally, the inclusion of representatives from different districts can be seen as a means to recruit and reward the politically active members of local elites, thus increasing the power base of the monarch. Although there is no doubt about the political nature of the NDC, the possibility that it contains a little bit more than that should not be totally discarded. The NDC can be seen as a preliminary experiment in allowing representatives of areas outside Kathmandu to voice their constituencies' interests in the forum of a national body and to partake in its activities. As such, this experiment could imply a certain degree of growing interest on the part of the palace to involve different districts in the country in the development effort. Their participation in the political-administrative process on a national level may produce a somewhat more significant input from still inactive or neglected parts of the country. This trend, initiated and guided by the palace, may produce a larger scope of development activity throughout the country as well as improved preparation of plans. However, before it becomes a reality numerous obstacles must be overcome, one of which is the reluctance of the Kathmandu-based elite to share its power and the benefits of development projects with others. It is still too early to evaluate the impact of the NDC in the above-mentioned context though it seems that thus far no significant progress has been accomplished. The NDC remains chiefly a debating forum that depends on material supplied to it by the bureaucracy, directives issued to it by the king, and the work of the substantially weakened NPC.[78]

Increased Powers in the Hands of the Ministry of Finance

In their analysis of the dynamics of Nepal's political system during the early 1960s, Joshi and Rose suggested the following—"One of the paradoxes of contemporary Nepali politics is that too obvious successes are as dangerous to the career of ministers as too obvious failures."[79] In 1976, the top echelons of the Ministry of Finance were approaching a stage where their efforts in the establishment of an effective organization could turn into a "dangerous success." The issue involved the political costs of attempts to improve the bureaucracy's performance and the dynamics of organizational reforms in a politicized bureaucracy.

During the early 1970s, the Ministry of Finance was assigned an increasingly larger scope of responsibilities. It was responsible for negotiating with aid-donor countries, and it was given legal authority over both programming and budgeting. Moreover, having been successful in the recruitment of a relatively large number of skilled personnel, the ministry intensified its efforts to introduce a more business-like approach in the preparation of development programs. By using the "power of the purse," the ministry assumed a central role in the preparation of annual plans and became involved in an increasingly careful scrutinization of data and programs presented by other ministries. These acts, combined with the recognized capacity of the Ministry of Finance, entailed unfavorable reactions on the part of practically all other ministries. The latter perceived these changes as an encroachment on their "territories." They considered the increased powers of Finance a threat to their own power and viewed the success of the latter a risk to their own capacity to perform in their areas of jurisdiction. To reduce this risk, the ministries resorted to tactics of foot-dragging in their interaction with Finance and complained to the palace of Finance's growing intervention in their respective affairs. Hence, the attempts of Finance to introduce additional changes in bureaucratic practices began to entail rising political costs. The Ministry of Finance stood to lose the cooperation of other ministries without which its position could be weakened. Moreover, with these rising political costs it stood to become a political liability in the eyes of the palace.

The palace, though interested in improving the bureaucracy's capacity to perform, neither sought to intensify power struggles within the administration nor wished to support any partcipant that could become a center of power. Accordingly, under circumstances of rising political costs, the Ministry of Finance could not expect its efforts to be backed by the crown. In fact, to secure the achievements it already accomplished, the ministry had to avoid any serious confrontations with other ministries over changes in planning practices and program implementation practices.

Ministerial Planning Units

During the years 1972 to 1976, different ministries established planning units within their organization. In this act the ministries complied with the formal decision to enlarge the planning apparatus (1968) but carried it out with considerable reservations. The idea to establish these units did not emanate from the ministries themselves, and they appeared to suspect the raison d'etre behind the function these units were expected to perform. The original proposals concerning the relationships between these planning units and the then-strong NPC seemed to increase the suspicions of the ministries. Unable to defy the order for the establishment of these units, the ministries sought to curb the role the former would be allowed to play in their respective organizations. The ministries were well aware of the power that could be achieved by a unit responsible for information processing and program evaluation. Accordingly, they sought to neutralize this unit from the outset. In most, if not all cases, the planning units were not seen in light of what they could contribute to improve the planning operation, implementation of practices, or evaluation of projects. Rather, they were seen as a new participant capable of becoming a strong contender in the competition over power within the organization. Consequently, these units were staffed with less than adequately skilled personnel and assigned responsibilities of drawing charts on the basis of information supplied to them by the top echelons of the organization and by project managers. The planning units were not given the manpower and means to verify data they received and were kept outside the information channels used by most units within the organization. Under these circumstances, the contribution of planning units to improved planning capabilities remained insignificant.

In summary, the changes introduced in the planning apparatus yielded a certain increase in planning capabilities. This increase was achieved incrementally and in a process characterized by frequent set-backs that emanated from the high degree of politicization within the bureaucracy. In recent years, the consolidation of political power in the center and an improved data basis enabled a relatively more rapid progress in the establishment of essential preconditions for effective planning. Nevertheless, the use of planning for development purposes remains largely dependent upon political considerations rather than on administrative considerations.

REFERENCES

1. N. Caiden and A. Wildavsky, Planning and Budgeting in Poor Countries, John Wiley & Sons., N.Y. 1974, pp. 168-169.

2. For discussion on the scope of this problem see, Ibid., pp. 167-238, 264-324; and A. Waterson, Development Planning: Lessons of Experience, John Hopkins University Press, Baltimore, 1969, pp. 61-169.

3. For a detailed discussion on political developments in Nepal during 1951-1960, see:

   a. Bhuwan Lal Joshi and Leo E. Rose Democratic Innovations in Nepal: A Case of Political Acculturation. University of California Press, Berkeley, 1966, pp. 83-394.

   b. R. S. Chauan, The Political Development in Nepal, 1950-1970. Associated Pub. House, New Delhi, 1971. pp. 112-165.

   c. Anirudha Gupta, Politics in Nepal. Allied Publishers Private, Bombay, 1964, pp. 51-272 and appendices II-V.

4. For the Rana's political system see:

   a. Padma Jung Bahadur Rana, ed. A. C. Mukherji, Life of Maharaj Sir Jung Bahadur of Nepal. Pioneer Press, Allahabad, 1909.

   b. D. R. Regmi, A Century of Family Autocracy in Nepal. Nepal: National Congress, Banaras, 1950.

5. Buch Commission, "Report on the Administrative Survey of Nepal." The M/S of External Affairs, Government of India, Delhi, April 1952, pp. 1-3.

6. For further details as well as a critical discussion on the role and contribution of the Buch Commission, see:

   a. S. D. Congal, "Survey of the Reports of Foreign Experts in the Administration of Nepal." Center of Economic Development and Administration, Tribhuvan University, Nepal, 1971, pp. 11-12, 65-76.

   b. M. K. Shrestha, A Handbook of Public Administration in Nepal. Nepal, Ministry of Panchayat Affairs, HMG, Kathmandu, 1965, p. 13.

7. N. L. Joshi, "Evolution of Public Administration in Nepal." CEDA, Tribhuvan University, Kathmandu, 1972, p. 18.

8. S. D. Congal, op. cit., pp. 12-13. See also N. L. Joshi, op. cit., pp. 24-27, and M. Goodall, "Development of Public Administration in Nepal," (Final Report), Kathmandu, 1962, pp. 1-10.

9. Officially prepared by the Ministry of Planning and development, this draft was actually the work of a foreign expert (on assignment to the Government of Nepal) and a few Nepali assistants.

10. Yadav, P. Pant, Planning Experiences in Nepal, Sahayogi Prakashan, Kathmandu, 1975, p. 23.

11. See Draft Five Year Plan, Government of Nepal, Kathmandu, 1956, pp. 4-8.

12. Resorting to such means would have required willingness on the part of the Council to risk an increase of tensions within the bureaucracy as well as more improper administrative practices to cope with it. Secondly, it would have required a strong commitment and willingness to perform the planning function diligently, regardless of the circumstances which made it difficult to do so. The Council, in its composition and in view of the prevailing circumstances, did not possess any of these inclinations. Furthermore, considering the fact that no one in the country knew how to cope with the function of planning, the Council could have chosen an easier way to satisfy formal aspects of its duty. It could use the data received from the ministries as the basis for planning, regardless of its quality and reliability. Such a pattern would have satisfied the formal necessity to produce a plan without causing serious infights among units of the bureaucracy. It would have made planning a futile exercise and plans mere documents, inconsequential in determining actual activity. This indeed was the case in later years|

13. B. L. Joshi and L. E. Rose, Democratic Innovations, pp. 234-235.

14. The vagueness which surrounded those projects, the lack of details with regard to budgetary allocation in the plan, and the two-year time limit which "coincided" with the proposed date for general elections, seemed to support the argument that the plan was mainly motivated by political considerations.

15. The parties in that coalition were: Nepali Congress, Nepali

National Congress, Praja Parished, and Gorkha Parished.

16. For some details on this plan, see T. Nath, The Nepalese Dilemma, Sterling, Pub. Pvt., New Delhi, 1975, pp. 78-79.

17. On events leading to the royal coup, see N. L. Joshi and L. E. Rose, Democratic Innovations in Nepal, pp. 374-392. Also see R. S. Chauhan, The Political Development in Nepal, pp. 109-166.

18. His Majesty's Government, National Planning Council, The Three Year Plan, (1962-1965). Kathmandu, 1963, pp. 49-53.

19. N. L. Joshi, "Evolution of Public Administration", p. 13.

20. For a detailed discussion on political development during 1960-1965, see:

   a. B. L. Joshi and L. E. Rose, Democratic Innovations in Nepal, op. cit., pp. 395-481, 485-517.

   b. R. S. Chauhan, The Political Development of Nepal, op. cit., pp. 167-267.

   c. For a review of political events, see T. Nath, The Nepalese Dilemma, op. cit., pp. 204-264.

21. For a description of the structure and functions of the Panchayat system, see:

   a. I. P. Kaphley, Fundamental Bases of Panchayat Democratic System, author's publication, Kathmandu, December 1967.

   b. U. N. Singh, Panchayat Democracy of Nepal in Theory and Practce. Department of Information, HMG, Kathmandu, 1972.

   c. For a penetrating critical analysis of the system, see R. Shaha, Nepali Politics: Retrospect and Prospect. Oxford University Press, Delhi, 1975, pp. 63-100.

22. Since appointments to top-level positions were made by the palace and since the government was the only major employer and entrepreneur, almost all actors preferred to work within the system, each waiting for his time to ride the 'merry-go-round.'

23. Since those effects became particularly evident during the later part of the Third Plan period, the theme will be discussed in

detail in Section 4 of this chapter and in the following chapter.

24. Some political leaders of the Nepali Congress Party, in exile in India, organized several resistance activities and raids across the border. Although this activity remained rather limited, it could not be stopped without India's cooperation. Indeed, it was only with the 1962 Sino-Indian dispute that India, in an effort to improve relations with Nepal, "advised" the rebels to discontinue their campaign. The advice was evidently accepted and the activity terminated.

25. The estimated number of officials dismissed during 1961 was put at more than 3,000. See Joshi and Rose, op. cit., p. 475.

26. Cited in T. Nath, The Nepalese Dilemma, op. cit., pp. 79-80.

27. K. P. Pradhan, Government and Administration and Local Government of the Kingdom of Nepal. Pradhan Pub., Kathmandu, 1969, p. 90.

28. HMG, NPC, The Three Year Plan, op. cit., p. 3.

29. Ibid., pp. 3-4.

30. Ibid., p. 6 and p. 25 (allocation of financial resources by sectors).

31. Ibid., pp. 97-99.

32. A. Wildavsky, "Why Planning Fails in Nepal," Administrative Science Quarterly, 17 (December 1972), p. 509.

33. B. B. Thapa, "Planning for Development in Nepal: A Perspective for 1965-1980." Doctoral dissertation, Claremont Graduate School 1967, pp. 18-19.

34. It evidently involved the decentralization of duties rather than the delegation of decision-making powers which were retained by the Palace Secretariat.

35. Joshi and Rose, Democratic Innovations, op. cit., pp. 476-477.

36. For a detailed account of those changes in the Council of Ministers, see Ibid., pp. 420-441.

37. HMG, Ministry of Economic Planning, The Third Plan (1965-1970). Kathmandu, 1965, pp. 3-4.

38. Based on several interviews with well-informed sources in Nepal. Kathmandu, February-April, 1976.

39. See T. Nath, *The Nepalese Dilemma*, op. cit., p. 82.

40. HMG, NPC, "The Third Plan," op. cit., pp. 11-18.

41. Ibid., pp. III-IV.

42. Nepal Rastra Bank *Annual Report 1969-1970.* Kathmandu, Nepal, pp. 3-5.

43. HMG, NPC, "Fourth Plan (1970-1975)," Kathmandu, 1970, pp. 32-33.

44. Ibid., p. 91 and p. 224.

45. Y. P. Pant, *Planning Experiences in Nepal*, op. cit., p. 40.

46. HMG, Ministry of Law and Justice, The Constitution of Nepal (as amended on January 27, 1967 by the first amendment of the constitution). Kathmandu, Nepal, 1967. Preamble, p. 1.

47. Ibid., Article 25 (p. 13).

48. Ibid., Article 26 (P. 14).

49. Ibid., Article 67A (p. 33).

50. Ibid., Article 86A (p. 46).

51. For some of the implications of the revaluation policy on the country's foreign trade situation, see T. Nath, *The Nepalese Dilemma*, op. cit., pp. 239-240, 409-411.

52. Ibid., p. 241.

53. For political events during the years 1966-1967 to 1975, see Ibid., pp. 239-302 and pp. 442-484.

54. Several of these changes actually took place at a somewhat later date, hence it was not immediately reflected in the role the NPC came to play during 1968 to 1971-72. Nevertheless, with the advance of time and particularly since 1972, it became apparent that the NPC was no longer among the more powerful units in the administration. This position was gradually taken by the Ministry

of Finance which absorbed various functions and some sub-units of the NPC into its own organization.

55. This measure is currently employed by the Ministry of Finance and it is yet to become fully operative. Difficulties in implementing it stem from the reluctance of ministries to cooperate in the matter, lack of effective control mechanisms that would make annual budgeting the decisive determinant in development activity and, dependency on flow of foreign aid for various development programs.

56. For changes in the budgeting procedures, see J. C. Beyer, *Budget Innovation in Developing Countries: The Experience of Nepal*. Praeger Publishers, New York, 1973, pp. 33-162.

57. The call for regional development planning was made by Harka Gurung who in 1972 became the vice-chairman of the NPC. See HMG, NPC, "Regional Development Planning for Nepal." Kathmandu, 1969.

58. HMG, NPC, "Fourth Plan," op. cit., p. 1.

59. Ibid., pp. 2-4.

60. Ibid., pp. 4-7.

61. Ibid., pp. 31-90 (agriculture); pp. 118-177 (transport and communications); pp. 191-209 (industry and trade); pp. 225-239 (power).

62. A. Wildavsky, "Why Planning Fails in Nepal," op. cit., p. 512.

63. Y. P. Pant, *Planning Experiences in Nepal*, op. cit., p. 67.

64. HMG, NPC, "Synopsis of the Mid-Term Appraisal Report of Fourth Plan." Kathmandu, 1972 (mimeo).

65. Based on interviews with high-level government officials, Kathmandu, January 1976-April 1976.

66. R. C. Malhotra, "Policy Analysis and Development in Nepal." Kathmandu, 1973 (processed), p. 6.

67. P. S. J. B. Rana, *Nepal's Fourth Plan: A Critique*. Yeti Pocket Book, Kathmandu, 1971, pp. 1-8.

68. See Nepal Press Report No. 260/1972 (June 10-11, 1972). Regmi Research Pvt., Kathmandu, 1972, pp. 1-2.

69. In September 1976 the King reorganized the NDC, increasing the number of its members. In addition to the above-mentioned members it now includes: the Chairman of the Rastriya Panchayat, the Chairman of the National Campaign Central Board, the Vice-Chancellor of the Tribhuvan University and, twenty-five District Panchayats Chairmen, instead of fourteen.

70. See A. Beenhakker, "An Inquiry into the Planning Function of Nepal." NPC and CEDA, Kathmandu, 1970 (internal report), pp. 1-24.

71. See J. C. Beyer, Budget Innovations, op. cit., pp. 29-51.

72. The Act, expected to be formalized in the 1976 Administrative Reform Act, was probably carried out, although a copy of the Act is not available. The Ministry nevertheless performed this function before formalization. See also the discussion in "Planning Processes."

73. R. C. Malhotra, "Policy Analysis and Development," op. cit., p. 26.

74. Ibid., pp. 16-20.

75. There are five Principal Royal Secretaries, each with its own secretariat, namely: 1) Principal Secretary to HM the King, 2) Principal Private Secretary to HM the King, 3) Principal Personal Secretary to HM the King, 4) Principal Press Secretary to HM the King, 5) Principal Military Secretary to HM the King.

76. D. R. Pandey, "Janch-bujh-Kendra, Ombudsman and Administrative Reform," Vashudha, Vol. 14, No. 1 (June-July, 1971), Kathmandu, pp. 33-39.

77. Based on interviews with government officials. Kathmandu, April-July, 1976.

78. The NDC held its fifth meeting from May 19 to 21, 1976. It deliberated on three papers presented to it by the NPC dealing with the outcome of the Fourth Plan. The Rising Nepal, May 20, 21, 22, 1976, Kathmandu.

79. Joshi and Rose, Democratic Innovations, op. cit., p. 439.

Chapter 4

THE POLITICS OF PLANNING IN THE TRANSPORT SECTOR

Introduction

Poverty was a characteristic feature of Nepal and a major obstacle in its efforts to change this condition. The dynamics of the development process were determined by this poverty and by the impact it had on the modes of operation of the participants. The mechanisms of this process and the uses made of them by the various participants were particularly prominent in the transport sector. The development operations in this sector were indeed a microcosm of Nepal's development efforts. The administrative and technical complexities involved in the establishment of a nationwide road network, the resources required for this undertaking and the multitude of participants and considerations involved in it, encompassed all major aspects pertaining to this effort.

The development process in the transport sector was composed of two broad and interrelated facets of interaction. First, it included the dynamics of Nepal's indigenous political-administrative system and the interaction that evolved between the components of which it was comprised (the internal arena). Second, it encompassed the interaction that evolved between Nepal's indigenous agencies and the agencies of the aid-donor countries which performed development tasks in Nepal (the external arena). These interactions were taking place in all phases of the undertaking—planning, preparation for implementation and the execution of transport projects. The driving forces behind these interactions encompasses a wide variety of political, administrative and economic interests of both the Nepali and the foreign participants. The uses made by these participants of the opportunities and constraints that Nepal's poverty presented to all, represented the differences in the operational capacities of these actors. Moreover, it accounted for the accomplishments and miscalculations witnessed in the different transport projects.

The stages of planning, preparation for implementation and the execution of these projects presented the participants with a certain set of operational problems that had to be resolved. Different actors found different solutions to these common problems. This variety in solutions represented the variety in the modes of operation of the participants as well as the variety in the patterns of interaction that evolved between these actors. The manner in which these interactions evolved and the results they produced exemplified the essence of Nepal's development process as well as its products. The operational problems that confronted the participants during the various stages of their development tasks covered a broad range of administrative functions. However, there were several functions that proved particularly problematic in the country's efforts to accomplish the development tasks. These included the functions of communication, coordination, the application of control mechanisms, and the management of personnel.

Each of these functions directly involved actions and resources that were of importance in the competition over power within the Nepali system. Hence, the fulfillment of these functions involved practices that were often incongruent with administrative criteria or development needs per se. Moreover, it introduced into the process a complexity that further increased the difficulties of various participants to perform their tasks. In the area of communication it involved deficiencies in the flow of information, problems related to the availability and reliability of data and difficulties in assessing the cause-effect relationships between actions and the progress achieved in the projects. In the area of applying control mechanisms it involved problems related to the financial administration of projects, widespread corruption and the difficulties emanating from the tendency to interpret administrative controls as part of a "power game" rather than as a legitimate means employed in the course of the administrative process. In the area of coordination it involved difficulties in attaining cooperation between units within the Nepali bureaucracy and difficulties inherent in managing large scale operations, as well as issues pertaining to cooperation between Nepali agencies and agencies of the aid donor countries. In the area of personnel management it involved the scarcity of skilled personnel, deficiencies in recruitment and training practices, widespread politicization in the promotion of officials, and difficulties in utilizing the available labor force. The impact of these factors on development operations was reflected in the manner in which the participants carried out both the planning and the implementation of the transport projects.

The process of establishing Nepal's national road network was a complex web of interactions that involved all the aforementioned components. The discussion below examines the manner in which this process evolved. This included the analysis of planning practices, the preparatory steps taken toward implementation and the execution of the transport projects. The analysis discusses modes of interaction between political and administrative factors, the role of economic considerations and the implications of Nepal's geographical setting. The present chapter and Chapter 5 deal with "planning" and "preparations for implementation" respectively and include the analysis of both the "internal arena" and the "external arena." Chapters 6 and 7 deal with "implementation"—Chapter 6 focuses on the "internal arena" whereas Chapter 7 examines the "external arena."

Setting Priorities—An A Priori Resolution?

The prevailing circumstances in Nepal in the early 1950s clearly pointed the necessity of establishing a nationwide transport network. Furthermore, the major components of the required network were also clearly identified. It was apparent that the regime, based in the center and wishing to enhance its governing authority throughout the periphery, would seek to establish direct contact with outer areas. In practice, it meant that Kathmandu would have to be connected with several central nodes in the country. Second, the country's topographical setting and the structure of the economy seemed to have "a priori" resolved the issue of identifying a few more major components of the required network. The traditional trade routes ran mainly in a north-south direction from Nepal to India. The Hills area which contained 60% of Nepal's population was poorly connected to the Terai area which produced approximately two-thirds of the country's agricultural output. It was obvious that if economic growth was to take place those two areas would have to be linked by means of land routes. Moreover, the necessity to transform centuries-old foot trails into a modern transport network seemed inescapable if increased economic interaction between different parts of the country was to become a reality. In brief, in the early 1950s Nepal was faced with the task of setting priorities among clearly required projects and to provide for their execution.

However, lacking the skilled personnel, the technical know-how and financial resources, it was also clearly understood that Nepal could neither plan nor execute any of those projects by herself. A constellation thus evolved in which the country had to depend almost totally on external assistance. The implications of this dependence were far more than the technical execution of projects. In fact, it meant that external interests were given the opportunity to play a central role in formulating Nepal's development effort. To Nepal it meant that in its drive to establish a road network, the country would have to consider not only its own particular interests but also those of the aid-donor countries. Moreover, it meant that on occasion the country would have to adjust its plans accordingly to offers received from external actors, and that some projects might have to be accepted even if they were only remotely connected to Nepal's more immediate needs.

Pressing Needs and Non-Refusable Offers—1952-1955

In the review of policies and programs in the transport sector (Chapter 2), it was pointed out that two road projects were among the major development undertakings carried out in the country in the early 1950s. Those programs were the construction of the Tribhuvan Rajpath and the Jogbani-Dharan Road. The Tribhuvan Highway was constructed by the Indian Army (1952-1956) and became the first major road in the country. It was (and still is) a highway of enormous economic importance for Nepal in general and Kathmandu in particular. This highway provided the first direct link between Kathmandu and the Indian border by means of a land route. Ever since its construction it has been the road on which the largest volume of goods moved in and out of Nepal. It brought about the substantial growth of Kathmandu and further enhanced the position of this particular area as the more developed part of the country and the undisputed political and economic center of Nepal. Indeed, if priorities had been set at that time (1952), the construction of such a road would have had to appear first on the list.

The Jogbani-Dharan road, on the other hand, had almost no viable economic justification to qualify it as one of the first projects constructed in the country. The road was built by the British army (1954-1958) and linked the foothills of Dharan with the Indian railhead opposite to Biratnagar in the Kosi zone. In itself this project neither served Nepal's immediate needs nor did it have a sufficiently sound basis to stimulate increased economic activity in the area through which it passed. It was a project which Nepal did not have to have and yet accepted with practically no reservations whatsoever.

The manner in which these two projects became part of Nepal's early efforts to establish a transport network seems, in retrospect, the beginning of a pattern of planning and implementation that gradually evolved in Nepal and blossomed in the transport sector in particular. In fact, both these projects came into existence more because of the external actors' interest in having them constructed than because of any strong Nepali initiative in the matter. In the case of the Tribhuvan Highway, India's initiative proved compatible with Nepal's needs, but it was the donor that initiated the offer just as it was Britain that came forward with a proposal to construct the Dharan road. Neither Britain nor India made their offers for "altruistic" reasons and Nepal accepted both regardless of this.

In making its offer, India most certainly considered the economic benefits that were to accrue to Nepal by having a direct link between Kathmandu and the Indian border. However, those considerations constituted only a small part of the raison d'etre behind the offer. Of much greater importance was India's own interest in having direct access to the political and economic heart of its neighbor to the north. To India, the establishment of such a road meant obtaining a means by which it could potentially increase its trade with Nepal and further utilize the gap between its industrial production and marketing capabilities and those of Nepal, all to India's advantage. Furthermore, such a modern road had substantial strategic importance, since it provided India with a greater potential to react to developments in Nepal and particularly those that involved Chinese moves in that area. In that respect, the Tribhuvan Highway should be seen in the context of the takeover of Tibet by China in 1950 and as part of the major road-building programs initiated by India in the Himalayas to compensate for the loss of Tibet as an autonomous buffer zone between India and China. Indeed, with Nepal suddenly becoming the immediate buffer zone, India apparently felt that it should have the means to enable it to move military forces swiftly into this area.

The peculiar position in which Nepal found itself in its efforts to have a road network constructed, and the pattern of trade-off between Nepal's interests and the donor's interests, were even more apparent in the case of the British project. The location of the project in this particular part of the country and the length of the alignment had little to do with Nepal's needs. It was a project based almost solely on Britain's interest in having better access to a Gurkha recruitment camp near Dharan and, presumably, a pay-off for the recruitment system. For Nepal however, this project was merely an addition to the road network, a British gift to the Gurkhas and a program carried out in an area that was not considered as top priority for development.

American Involvement in a Road Project, 1955

In 1955 a third aid donor country became involved in the construction of roads in Nepal. Once again the initiative to establish this road emanated from the donor and the recipient accepted the offer with little reservations. During this year, the US (or more accurately the USOM mission in Kathmandu) seemed to realize that American aid programs in Nepal were distributed over a relatively large number of fields but lacked political visibility as well as economic impact. Moreover, it was felt that the yield of these programs was somewhat incompatible with the ideas of the "American philosophy" of foreign aid—e.g. establishment of a democratic way of life, widening the base of the government, and increasing mass participation in government by changing attitude about the responsibilities of individuals and villages in national life.[1] The target therefore was to find a new project which would have the potential of being a model for all Nepal. Such a project was found in the Rapti Valley and Nepal accepted the American offer to undertake it. There were, as Mihaly pointed out, compelling reasons for carrying it out from both the American point of view and Nepal's interests.

> The Rapti Valley bordered on crowded, land hungry regions of Nepal and could therefore serve to offer land to the landless. Because a fresh start could be made, the Rapti was a potential model for all Nepal—a vast demonstration project in which equitable land distribution could be coupled with more productive farming methods, better health and social services. Moreover, because the Rapti lay near the main road from Kathmandu to the Indian border, a cash economy could be introduced.[2]

The project was indeed of a substantial magnitude. It involved a mechanized land-clearing operation, the establishment of eighteen cooperative societies, a model health center, and an experimental farm. It also included a craft training project, the construction of a sawmill, and the building of a road that would cross the valley from Hetaura to Narayangha.

By 1959, due to a variety of reasons it became apparent that the project fell far short of its considerable potential. In fact, the road project was among the most visible and lasting projects of the entire scheme. However, its use as a catalyst for economic growth was greatly diminished by the collapse of the larger project. Nepal therefore had an additional road which connected two points in the country but which proved to be neither the product of adequate planning nor a sufficient condition for generating economic growth in the area.

Early Steps in Planning—1956-1960

The need for a better, more comprehensive development plan was recognized by the Nepali authorities in the country's First Five Year Plan (1956). The plan, which did not provide any specific targets in most sectors of the economy, proposed nonetheless that during the five-year period "300 miles of metalled road, 300 miles of fairweather road and 300 miles of nine-foot jeepable track will be constructed." It suggested that "During each subsequent Five Year Plan the fairweather road track constructed during the preceding years will be converted to metalled road, and a new 600 miles of fairweather road and 600 miles of track will be constructed."[3] Considering the capabilities that existed in Nepal at that time, and the slim potential for the successful achievement of somewhat more modest objectives, this proposal could not be anything but a mere fantasy and an illustration of an unrealistic planning process. Still, by 1958, the target of constructing 900 miles of road in five years was adopted by the newly-formed Regional Transport Organization (RTO). Planning in this tripartite body—a joint venture of the United States, India, and Nepal—was as deficient as Nepal's Five Year Plan. No surveys were made before the target of 900 miles was written into the agreement: it was an educated guess made by the Head of the US Technical Cooperation Mission in Nepal under the pressure of time as funds suddenly became available and there was a rush to reach an agreement on how to use them.[4]

In fact, the dynamics of the planning process in the transport sector of Nepal were clearly reflected in the manner in which the RTO was established as well as in the structure of this organization and the plan which it adopted. The Nepali government realized that transport projects were an effective means to attract most needed investments. Aid-donor countries, realizing the political visibility of road projects and the length of time these projects enabled the donors to operate in the country, were willing to invest in this particular sector. Hence, when the US made its offer to establish an organization that would construct 900 miles of roads, Nepal immediately accepted the offer. The US made its offer not so much because it knew how the RTO could most benefit Nepal's development effort; rather, the offer was made because the US was interested in increasing its role in the country and because funds necessary to make such a move became available in its Asian Economic Development Fund. In that respect both the donor and the recipient shared a mutual interest—they wanted to have an agreement signed and the funds committed without delay. The problems of clarifying details of the operation, specific targets of the organization, and its operational procedures were left for a later date. Furthermore, the tripartite form of the RTO, which included India as the third member of the organization, came about because it was required in the AEDF regulations, and not because it was considered the most suitable form for executing road projects in Nepal.[5]

Conflicts of interest between the US and India, and between Nepal and India produced constant and serious disagreements among the members of the organization. India was interested in roads leading to its border and to other Indian projects in Nepal, whereas the US was looking for projects that would be more closely associated with the American assistance efforts. Nepal, overwhelmed by the magnitude of the program, proved incapable of setting priorities among projects and areas it wanted developed. Furthermore, American and Indian personnel were engaged in a destructive competition over whose methods in road construction would prevail and over the proper way to administer these projects. An additional element of ineffectiveness emanated from the insistence of the US on working with the Nepali bureaucracy whom it mistakenly considered capable of providing the complex administrative support required for the projects.

Plagued by political conflicts and administrative inefficiency, the RTO first cut down its target from the original 900 miles to 300 miles. Then, by June 1962 the entire venture was abandoned with only negligible accomplishments to its credit. Four years of this joint operation saw the construction of 148 miles of jeepable track (no single stretch of which was long enough to connect two towns) the pavement of 24 miles of existing road, grading 193 miles, and laying gravel on 41 miles.[6] Evidently, the RTO's operations did not provide Nepal with one additional usable road!

The factors that determined both the RTO's targets and its activities were clearly reflected in two particular projects, the Kathmandu-Trisuli Road and the Dhangari-Dandeldhura Road. The inclusion of the first project in the organization's scheme was, most probably, done for reasons of formal bureaucratic requirements, being a project carried out by one of the members of the RTO (India) at the time the organization was in operation. In practice it was an Indian project, an integral part of the Trisuli hydroelectric scheme (another Indian project), and one which was carried out mostly by India although some American engineers and equipment were employed in the project. India considered the Trisuli hydroelectric scheme an undertaking of substantial importance. Accordingly, it did not allow either the establishment of the RTO or its dissolution to interfere with its completion. Indeed, once the RTO was dissolved, India completed the road project with its own resources (1963).

The Dhangari-Dandeldhura road was one of the largest projects expected to be carried out by the RTO. Yet, in 1962, when the organization was dissolved, progress on this project amounted to a preliminary reconnaissance survey of the proposed alignment and the construction of 30 miles of two-foot track, as compared with a target of 91 miles of motorable road.[7] Technical and administrative difficulties in executing this project were enormous, but the problems began much earlier and were rooted in the factors that led to the decision to launch this project. This decision was an outcome of American and Nepali interests coinciding to start a road project in the far western part of the country. To the US it meant having a project that could be identified as "American" and one which would be carried out in an area where no other external assistance activity was underway. Moreover, it seemingly symbolized, in the eyes of some American officials in Kathmandu, the essence of a proper foreign aid program since it was to be executed in one of the least developed areas in the country. ("Proper foreign aid" presumably meant aid given where it was most needed and where it met the recipient's objectives to alleviate harsh economic conditions.)

To Nepal, or more accurately, to King Mahendra, the launching of a sizable project in this particular part of the country was an act of internal political significance as well as a move to counter-balance developments on the Indian front. In the internal arena it presumably symbolized the regime's concern for equal distribution of the benefits of aid and development throughout the country. On the Indian front it was intended to counterbalance the potential effects of India's road building program directly across the border. Thus, Kathmandu chose an alignment that did not link up with Indian roads or the Indian railhead in Tanakpur and the areas surrounding it.[8] In other words, although the Dandeldhura-Mahendranagar alignment had a much greater economic justification than the Dhangari-Dandeldhura, the latter was chosen because it did not link up with the Indian road network.

In 1962 Nepal had only three major road projects completed and none of them was the result of a carefully prepared Nepali plan. The planning process proved deficient and the projects undertaken were based on the availability of external offers. Under the then prevailing circumstances Nepal could not realistically expect the situation to change rapidly or drastically. However, were Nepal to assume a more active role in the setting of priorities among projects and areas of development, it could exploit more effectively the external interests in the transport sector. Subsequently, this strategy was adopted by King Mahendra and gained an impressive measure of success.

New Participants, Similar Principles—China 1961

China's aid program in Nepal began in 1956 and remained relatively limited up to the early 1960s. It included grants which were offered under favorable terms, and offers of technical assistance projects which did not seem to have any strings attached to them. There was little doubt that Nepal needed whatever aid it could get and that Beijing made its offers primarily to create a more sympathetic attitude toward China in Nepal. Chinese aid during that period included, among other things, a $100,000 contribution for the construction of a 55-bed hostel for a Buddhist school outside Kathmandu.[9]

A significant turning point in the scope and form of Chinese aid to Nepal came in 1961. On September 25, 1961, King Mahendra began a 17-day state visit to China and Outer Mongolia. The stated purpose of the visit was the signing of a boundary agreement between the two countries. The agreement, signed on October 5, 1961, seemed to satisfy most of Nepal's claims over the areas under dispute and the King obviously welcomed it. However, on the day before his departure, the Chinese presented King Mahendra with a draft road agreement, a subject that had not been on the agenda and was not raised by either side during the entire visit. Moreover, the presentation was done in such terms as to imply that the implementation of the boundary treaty depended upon a favorable response on the road issue.[10]

Apparently a package settlement on boundary matters suddenly became a "package deal" that included an agreement on the construction of a road between Kathmandu and Tibet. Unwilling to reject the entire deal, Mahendra consented to the inclusion of the road agreement as part of the deal and had his Prime Minister sign it.[11]

The potential importance of the road to Beijing was substantial, judging by the urgency with which China approached the project. Moreover, favorable terms to Nepal were incorporated in the treaty despite a severe economic crisis which placed a heavy strain on China's limited foreign exchange reserves. Nepal, on the other hand, was about to have a new road constructed in its territory, another project which it had not planned.

China's interest in a road linking Tibet with Nepal was partly based on a desire to expand its influence in this area and partly on the then pressing need to solve difficulties in supplying the Chinese military establishment in Tibet. The roads into Tibet from Szechuan (on the east) and Tsinghai (on the northeast) traversed extremely difficult terrain, were expensive to maintain, and were subject to sabotage and blockade by the Tibetan rebels, the Khampas. The road from Singkiang (on the northwest), although easier and safer, crossed the Askai Chin plateau which was then an object of bitter dispute between India and China. Furthermore, India's ban on trade in strategical goods with Tibet (1960) eliminated China's primary source of supply for many commodities. A road linking Tibet with Kathmandu had therefore the potential of solving this multi faceted problem, since Nepal, unlike Sikkim and Bhutan, was not included in the Indian trade blockade. Rice from the surplus production areas in Nepal (the Terai) and manufactured goods could therefore be imported from Nepal into Tibet more easily, cheaply, and quickly than directly from China.[12] Militarily, this proposed road also had the potential of providing China with a suitable access across the Himalaya into the subcontinent (via Kathmandu and down the Indian-built road in the Terai to India).[13]

In March 1962 Chinese technicians began to survey the area and proposed two alternative routes, one through the Rasua Garhi pass and the other via Kodari, the lowest all-weather passes in the entire Himalaya range. China preferred the Rasua route which was shorter by nearly 20 kilometers and traversed easier terrain but accepted Nepalese demands that the Kodari route would be chosen. This appeared to be the extent of maneuvering room that Nepal was given in planning this road project.

Nepal Takes the Initiative—The East-West Highway

During the years 1951 to 1959, developments in the transport sector clearly indicated that road projects were attracting aid donor countries. It was also evident that Nepal played a rather limited role in initiating development programs in this sector.[14] Toward the end of the First Five Year Plan, the regime recognized the necessity for Nepal to assume a more active role in the formulation of development priorities. Second, it realized that the competition between prospective aid donors could be enhanced and exploited to the benefit of the recipient. These two principles were translated into practice in a grand Nepali initiative taken in the late 1950s in the transport sector—the construction of a new highway that would traverse the country from east to west.

In the late 1950s, the establishment of a cross-country highway could not be justified in economic terms.[15] However, from the regime's perspective, this program could be of considerable political importance. It could provide direct contact between all parts of the country and enable the center to penetrate into different regions of the polity. Moreover, it could appeal to nationalist feelings since it would provide local citizens with a means to traverse the country without having to use India's transport network. Nevertheless, it was obvious that Nepal would not be able to carry out this project by itself. Hence, in late 1958 and early 1959, requests were made to India and the United States to undertake the project. These requests were turned down by both who argued that such a road was not yet economically justifiable. Considering the ratio of projected economic utility of the road to the magnitude of the investment required, this was indeed a valid argument. Moreover, upon considering its own particular interests in the area, India could not fail to see the strategic significance of such a road—a means that could lead to a reduced dependence of Nepal on India. Obviously, India was less than eager to see this happen, much less assist in making it a reality.

The failure to obtain assistance from the United States and India seriously reduced the prospects of starting the project. However, by mid-1959, the situation seemed to be changing as a new participant appeared on the scene. The USSR, eagerly seeking Mahendra's friendship, proposed a Soviet aid program in Nepal. On April 29, 1959, following a series of deliberations, an agreement on economic and technical assistance was signed by the two countries. The agreement stipulated that the USSR would give "free of charge economic and technical assistance worth $7.5 million to assist Nepalese state organizations in the construction of a hydro-electric power project, a sugar factory, a cigarette factory, and a survey for an east-west road not exceeding 1,050 km in length."[16] The Soviet Union approached this aid program on the basis of certain long-term objectives which characterized Soviet aid to other underdeveloped countries: the support of the recipient's non-alignment policy, the introduction of an alternative to economic dependence on the West, and the strengthening of the state through the construction of state enterprises.[17] The first Soviet aid was therefore structured to serve the interests of the donor who wanted to gain better access to the country and to provide the recipient with assistance where it was most clearly needed. The electric power project and the sugar and cigarette factories could be beneficial to Nepal since there was a great demand for these commodities in the country. The road-survey project could be seen as a politically important investment: it could be presented as a sign of the "sincere Soviet willingness" to respond favorably to Nepali requests and assist in the development effort. This latter aspect of Soviet aid gains increased importance when set against the refusals of both the US and India to undertake the project.

Implementation of the Soviet projects began in late 1959 and early 1960. However, numerous difficulties hampered the progress of those projects and relations between the Soviet personnel and the Nepalese deteriorated. In 1961 it became apparent that HMG was mainly interested in the road project which meant that the other projects were given only secondary importance. In fact, following the royal coup, the regime attempted to draw the USSR into the construction of the East-West Highway. These efforts proved to be fruitless and the Soviet Union not only declined to accept HMG's invitation but also refused to complete the second half of the survey.[18] Apparently the Russians concluded that the best course of action would be to disassociate themselves entirely from the project. Moreover, when seen in the context of the deteriorating relations between the USSR and China and the changes in Soviet-Indian relations, the Soviet's interests in this particular area diminished. Indeed, when Indo-Russian relations improved in the mid-1960s, the USSR evidently agreed to resume its participation in the construction of the East-West Highway. Nevertheless, in

1961/62 Nepal found itself with a project it wanted to develop but with no donor showing the least interest in it.

King Mahendra, who considered the project far too important to be abandoned because of lack of external participation, thereupon called his countrymen to undertake it themselves. The call, which was made prior to the launching of the Second Plan, emphasized the potential importance of this road to Nepal. It was portrayed as the lifeline of the country and implicitly presented as a sign of Nepal's growing independence from external sources, particularly India. Reportedly, the call generated substantial enthusiasm among the Nepalese and the project was immediately incorporated in the Second Plan.[19] Construction of the road was expected to be carried out by voluntary work by all Nepali citizens. Additionally, the move, which coincided with the dissolution of the RTO, was supplemented by some new organizational arrangements in the transport sector. A high level committee—the East-West Highway Committee—consisting of all cabinet ministers with King Mahendra himself as its chairman was formed "to direct this active cooperation of the local people in the construction of the highway."[20] Secondly, HMG's Roads Department, which up to this date had undertaken only minor operations, was assigned new responsibilities and designated as the major organization in the transport sector. Seemingly, Nepal decided to take a substantially more active role in the transport sector beginning with the Second Plan.

The Second Development Plan

Under the Second Development Plan, the road-construction program was structured around certain particular projects, the major one being the East-West Highway. Of the 724 miles of fair weather roads targeted for construction during the plan period, 530 miles were on the East-West Highway alignment which corresponded to the part of the alignment surveyed by the Russians. Other major projects were the Kathmandu-Kodari road (China), the Kathmandu-Janakpur road (an all-Nepali road), the Sunauli-Pokhara road (India), and the completion of the Kathmandu-Trisuli road (India). Some road improvement operations on the Tribhuvan Highway and on roads in the Kathmandu Valley were also expected to be completed during these three years.[21]

The targets set in the Second Plan reflected the high expectations among many Nepalese and a strong sense of confidence in the country's ability to accomplish them. However, developments soon proved disappointing, in particular the Kathmandu-Janakpur road and the East-West Highway. On that subject one Nepali commentator wrote that "because of poor planning and ineffective supervision, a great deal of popular enthusiasm, efforts and resources were wasted."[22] Indeed, 18 months after the highway project was launched it became apparent that Nepal was incapable of carrying it out. Abandoning the project, however, was evidently inconceivable to King Mahendra. He turned once again to external sources in search of an "exploitable" constellation that would permit the continuation of the project. This was found in the context of relationships between Nepal's neighbors—India to the south and China to the north, particularly in view of the 1962 Sino-Indian dispute.[23] Both countries sought to maintain a close relationship with Nepal and HMG was able to obtain come concessions as well as additional aid from both.[24]

In late 1963 Nepal approached China on the Janakpur to Biratnagar sector of the East-West Highway. Beijing expressed its willingness to undertake the project and an agreement was actually signed between the two countries.[25] At that point India came forth with a proposal to construct this portion of the highway as well as the section eastward from Biratnagar to Kankharbhitta. To India, the idea of having Chinese technicians posted so close to its border was seen as a threat that could not be allowed to become a fact. To Nepal, the move proved highly successful since India offered rather generous terms to undertake the project, while China agreed to undertake another road project in the Hills in place of the East-West Highway—a road between Kathmandu and Pokhara in western Nepal.

Kathmandu's success in playing off China against India also enabled Nepal to approach the Soviet Union again on the East-West Highway project. This time, backed by India's commitment to the project and having the Hetaura-Narayangha section already completed by the USA, Nepal's efforts succeeded. In 1965, the USSR agreed to undertake the road from Janakpur to Simra—the Dhalkewar-Pathlaiya section. Exhibiting impressive political skills, King Mahendra achieved what he had been aiming at all along—the construction of a highway that would traverse Nepal.[26] The series of agreements signed with India and the Soviet Union, together with the already completed section between Hetaura and Narayangha, accounted for about half of the projected total length of the East-West Highway. Later, in 1967-1968, an agreement was signed with Britain to construct the section from Narayanha to Butwal and in 1972 India agreed to construct the remaining Butwal-Kohalpur sector. Thus, in a masterly display of political acumen, King Mahendra extracted commitments from rather reluctant aid-donors for the construction of the East-West Highway.

The Master Plan—Or, How Much Ex Post Facto Planning is Inevitable?

The success achieved by Nepal on the East-West Highway did not solve the country's difficulties in preparing a plan for the transport sector. On the contrary, it was now clear that Nepal lacked the necessary skills and resources to perform such a planning task. Consequently, in 1964 Nepal asked the World Bank for assistance. A team of experts arrived in Kathmandu and was assigned the task of preparing a Five Year Transport Sector Plan for inclusion in Nepal's Third Five Year Plan. Due to the then prevailing circumstances in the country, in particularly the lack of data on the economy, the Bank's team decided to approach the matter from a much broader perspective: with the blessing of HMG, the team undertook to produce a Twenty Year Master Plan for the Transport Sector and submitted its final product in June 1965. The Master Plan contained a large number of road projects for various parts of the country.[27] It also included a proposal for the establishment of a new organization, the Road Authority, which would coordinate and supervise all road projects in the country.[28] In preparing the Master Plan, the World Bank's team mainly dealt with the economic and technical aspects of transportation and suggested some administrative measures that would enable the implementation of the proposed projects. However, aware of the role political considerations often played in the transport sector, the team cautiously noted that "The Master Plan . . . is in effect a perspective plan giving a general idea of the goals to be reached in, say, a thirty year period of time . . . . The Master Plan in itself is only a beginning. Starting from this base, transport programs can be evolved for each successive planning period."[29] With this Master Plan, the transport sector became the first in the country to possess a long-range plan. What remained to be seen was the extent to which this plan would be used in Nepal's development effort in the transport sector.

Thus far the argument has been made that predetermined plans were not major factors in development activity. Rather, this role was played by internal political-administrative factors and external interests, as development activity was largely determined by the circumstances that prevailed in the country at any given point of time. The case of the Master Plan's proposal for a primary road network presents the broadest and most significant example of this phenomenon. A superficial comparison between the primary road network, as proposed in the Master Plan, and the existing primary network reveals that the planned and the actual networks are practically identical. However, a closer examination clearly indicates that this similarity had little to do with the planning process per se.

The primary road network as it exists today consists of the following major components: the Mahendra Highway (East-West Highway), the Tribhuvan Highway from Kathmandu to Raxaul, the Arniko Highway from Kathmandu to Kodari, the Prithvi Highway from Kathmandu to Pokhara, and the Sidhartha Highway from Sunauli to Pokhara. All of these projects were presented to the planners as a fait accompli even though the construction of some had not been started by the time the Master Plan was finalized by HMG in 1966-1967. The shape of the primary road network as proposed in the Master Plan, thus, was ex post facto planning. Furthermore, it actually required that the planners propose a plan on the basis of this predetermined primary network.

The Third Plan, 1965-1970

A review of the program for the transport sector in the Third Five Year Plan reveals that in the mid-1960s Nepal did not yet enjoy the luxury of making such decisions on the basis of economic considerations. The projects planned for implementation during the Third Plan period were, with only one exception, those decided upon earlier, including the Kathmandu-Pokhra (China), the Kathmandu-Kodari (China), and the Sunauli-Pokhara (India) projects. The construction of the East-West Highway was expected to continue and to include the section from Narayangha to Butwal. One new project was proposed for the Far Western region—a road between Mahendranagar and Doti.

A new element introduced into this periodic plan was aid to Panchayats for the construction of local roads. The National Transport Organization, established prior to the launching of the plan, was assigned the responsibility of coordinating the operations of the ropeway and railway in the Bagmati and Narayani zones.[30] Apparently, the Master Plan had little to do with projects included in the Third Plan. China completed the Kathmandu-Kodari road in 1967 and began to work on the Kathmandu-Pokhara road. As China objected to constructing a road that would connect Kathmandu to the major Indian project at Trisuli, the Chinese chose a road alignment along the Naubise rather than along the alignment proposed in the plan, i.e. via Trisuli. Nepal accepted the change with no evident objections even though it made little sense economically. In the meantime, India, which was interested in roads that would link its border areas with inner Nepal, was engaged in the construction of the Sunauli-Pokhara Highway (the Sidhartha Highway) in west-central Nepal. In 1968 Nepal persuaded Britain to undertake the construction of the Narayangha-Butwal section of the Mahendra Highway. The manner in which Britain was drawn into the project was probably similar to the way in which Soviet involvement was secured. Britain, interested in maintaining friendly relationships with Nepal but with only a small aid program in the country, agreed to join the project. The argument used was that as all other major aid-donor countries were participating in it and that the regime considered this highway a project far too important for Britain not to take part in it. Furthermore, Nepal also argued that in view of a century of Gurkha recruitment to the British army, Britain owed Nepal a "debt" and this project was a particularly good opportunity to repay it.[31]

The Third Plan included one new major project which was proposed in the Master Plan, a fair-weather road between Mahendranagar and Doti. In terms of projected economic returns, the planners recommended that the best route would be from Dhandehdura to Mahendranagar.[32] However, a reconnaissance survey conducted in 1967 resulted in a different proposal—an alignment between Dhangari and Dandeldhura.[33] The latter was chosen by the regime, even though there was almost no economic justification for this particular alignment.[34] The decision was based on two particular factors: first, it was felt that the Far Western region had to have at least one large development project underway and, second, for such a project to enjoy a sound financial backing, it was felt that external assistance would have to be obtained. The Dhangari-Dandeldhura project had once been projected as an American assisted project, and US participation in it could be revived. Indeed, in 1968, following a series of deliberations between HMG and USAID the latter became involved in what has become an interminable project.[35]

In sum, the recommendations made by the Master Plan were not incorporated into the Third Plan. The majority of the road projects launched during the Third Plan were decided upon on the basis of political considerations rather than on the basis of the predominantly economic approach emphasized in the Master Plan.

Local Roads and Political Participation

Prior to the launching of the Third Plan and with the exception of the capital city, Kathmandu, no serious efforts were made to establish networks of local roads. With the inauguration of the Panchayat system, the attitude toward this segment of the transport sector changed and local elites were called upon to propose and initiate development projects in their respective villages. The change was part of the regime's drive to strengthen the newly established political structure and in congruence with the widely publicized principles of the Panchayat system, i.e. popular political participation and greater involvement of all citizens in the country's development efforts. In adopting this change in attitude, the regime assumed that its support to initiatives at the local level would enable it to deepen its penetration into different regimes of the country. Similarly, it assumed that by way of budget allocations and the extension of bureaucratic services to local programs, the dependence of local elites upon the center would increase. On their part, the local elites were relatively quick to adapt to the call from the center. They recognized the advantages inherent in budgetary support from the center and they were well aware of the limited capacity of the center to monitor progress in local projects. Furthermore, with the center's emphasis on the transport sector and in view of the considerable involvement of aid donors in road projects, local elites did not ignore the prospective possibility of attracting foreign-currency-supported projects to their respective villages. Consequently, beginning with the Third Plan and extended well into the Fourth Plan, the number of project proposals initiated at the local level rose sharply and included a significantly large number of proposals for the construction of local roads.[36]

The local elites did not limit their efforts to the preparation of proposals. Rather, they became actively involved in mobilizing the local population to launch local projects. In road projects, these initiatives entailed the recruitment of "volunteers" that embarked on the construction of a few miles of fair weather roads or the widening of a foot path. It was asserted that once the local population made its contribution to the development effort, the regime would be more willing to support this local initiative and allocate it additional resources. Kathmandu accepted this mode of operation but extended financial and technical support only to those projects it considered important. Subsequently, this led to a gradual growth in the number of local projects supported by the center and to a gradual increase in the participation of the center in local projects.

Toward Improved Planning: The Fourth Plan

Compared to its predecessor, the Fourth Five Year Plan (1970-1975) relied more heavily on the Master Plan and emphasized the completion of projects already underway as well as the construction of north-south alignments.[37] It recognized that without north-south alignments and feeder roads that would link with the primary road network, the utility of the latter would be greatly reduced. Accordingly, it projected the construction of the following roads—Nepalgang-Surkhet, Putlikhet-Baglung, Trisuli-Dhunche and Dharran-Dhankuta.[38] It also included the construction of some inter-district and intra-district roads such as the Gorkha-Pritvi Highway, the Krishnagar-Mahendra Highway, the Dang-Mahendra

Highway and the Rajbiraj-Mahendra Highway.[39] Furthermore, on the basis of the arguments presented in the Master Plan, the central authorities sought assistance in the planning of the country's secondary road network. HMG requested the UNDP for this assistance and, by the time the Fourth Plan was launched, the consulting agency assigned by the UNDP began its project.[40] This project included a reconnaissance survey of the secondary network, feasibility studies of 18 roads and a feasibility study of the Kathmandu-Raxaul Corridor.[41] The UNDP also provided technical assistance to the Roads Department and the Ministry of Works and Transport.[42] In 1974, the consulting agency submitted its reports and a follow-up study was then conducted by the UNDP on the 18 road projects and the reorganization of the Roads Department.[43] By the end of the Fourth Plan, Nepal had the general framework of its secondary road network planned and thus possessed a program for the entire road network in the country.

During the course of the Fourth Plan, most of the projects in the primary road network were either completed or nearing completion. One exception was the East-West Highway. India began the construction of the western portion of this highway, from Butwal to Kohalpur in 1972, but the Nepali agencies that were expected to construct the final leg from Kohalpur to Bhanbasa did not progress beyond the stage of preliminary planning. Among the north-south roads planned for construction, only limited progress was achieved in two projects—the Nepalgang-Surkhet road and the Dharan-Dhankuta road. The first was a project undertaken by the Roads Department and the second was under construction with British assistance. The high priority given to these two projects emanated from the regime's decision to enhance development by way of establishing four Development Regions throughout the country. The headquarters for the Eastern Region were located in Dhankuta and the headquarters for the Far Western Region were in Surkhet (the other two were in Kathmandu for the Central Region and in Pokhara for the Western Region).

Over the past 25 years, planning in the transport sector developed in three stages. At first, it was a mere administrative ritual that lacked any bearings on actual development activity. During this stage, which lasted from 1951 to 1959, the planning of transport projects was subservient to external offers pledged to particular road projects. Then, during 1960 to approximately 1968, it evolved into a means used by the regime to enhance and exploit competition between prospective aid donor countries. During this second stage Nepal attempted to establish a more sound data basis for future planning activities and began to take a more active role in leading aid donors into projects it considered important. The third stage began toward the end of the Third Plan and involved a slow but discernible movement toward the use of planning as a framework for action. Internal and external political considerations continue to determine which projects will actually be implemented, but in comparison to earlier stages the planning of these projects also rests upon a somewhat more sound economic basis.

REFERENCES

1. For a description of US assistance to Nepal during the early 1950s see: E. B. Mihaly, Foreign Aid and Politics in Nepal: A Case Study, Oxford University Press, London, 1965, pp. 69-72.

2. E. B. Mihaly, Ibid., p. 76.

3. Government of Nepal. "Draft Five Year Plan," Kathmandu, 1956, p. 40.

4. Mihaly, op. cit., p. 83.

5. The Asian Economic Development Fund was a regional fund created by the U.S. for the purpose of financing projects which would assist two or more countries in the region. The fact that the RTO was expected to construct some roads that would link Nepalese towns with India brought the latter into the organization, thus satisfying that stipulation. In this case therefore the administrative regulations of an external assistance fund came to determine the structure of the RTO. It soon became evident that this uneasy and ill-devised partnership actually prevented the organization from attaining its goals.

6. HMG, NPC, "The Three Year Plan (1962-1964)." Kathmandu, pp. 151-152.

7. R. S. J. B. Rana, An Economic Study of the Area Around the Alignment of the Dhangari-Dandeldhura Road, Nepal. Kathmandu, CEDA, 1971.

8. Based on interviews with well-informed sources in Nepal. Kathmandu, May 1976.

9. On Chinese assistance programs in Nepal during 1956-1963 see: Mihaly, Foreign Aid and Politics in Nepal, op. cit., pp. 152-156. See, also: L. E. Rose, Nepal—Strategy for Survival. Oxford University Press, Delhi, 1971, pp. 210-218.

10. L. E. Rose, Ibid., p. 239.

11. Since Mahendra knew India would strongly object to this agreement, he preferred not to sign it himself as he had the border agreement. Under the terms of the agreement, China promised monetary aid to Nepal amounting to approximately $10,000,000 for construction work and also agreed to supply the necessary experts, technicians, and equipment. Moreover, Nepal was given, at least theoretically, the final voice in the implementation of the project.

12. Ibid., p. 240.

13. E. von Pfister, "Highway Projects on China's Southern Border," Vasudha, Vol. xiv, no. 4 (Nov.-Dec. 1971), pp. 11-12. (Reproduced from Aussen Politik, German Foreign Affairs Review, Vol. 21, March 1970.)

14. Though an all-Nepali road (to Tibet) was considered by T. P. Acharya's government in 1956 nothing of substance came out of it and Nepal's role in this sector remained limited.

15. HMG, NPC, "The Three Year Plan," op. cit., p. 157.

16. E. B. Mihaly, Foreign Aid and Politics in Nepal, op. cit., p. 98.

17. Ibid., p. 157.

18. Ibid., p. 161.

19. Based on interviews with many Nepalese officials and citizens, Kathmandu, December 1975 - July 1976.

20. HMG, NPC, "The Three Year Plan," op. cit., p. 155.

21. For a detailed account of planned projects in the transport sector during the Second Plan see: Ibid., pp. 154-160.

22. R. Shaha, "Foreign Policy" in P. S. J. B. Rana and K. P. Malla (eds.), Nepal in Perspective, CEDA, Kathmandu, 1973, p. 255.

23. L. E. Rose, Strategy for Survival, op. cit., p. 248.

24. For a discussion on relationships between Nepal and China and Nepal and India during 1963-1971 see: L. E. Rose, Ibid., pp. 250-276, and Ramakant, Nepal—China and India, Shakti Malik Pub., New Delhi, 1976, pp. 200-251.

25. R. Shaha, "Foreign Policy," op. cit., p. 255.

26. Originally, King Mahendra wanted an East-West Highway in the Hills rather than in Terai. However, since India was willing to provide assistance only on the condition that the alignment would pass through the Terai (i.e. closer to the Indian border) Mahendra evidently changed his plan and accepted the alternative proposal.

27. For a detailed description of the Master Plan's

recommendations on transport projects in various parts of the country, see: IBRD, A National Transport System for Nepal, Washington, D.C., June 1965. Vol. II (supporting Technical Papers and Discussions), pp. 105-145.

28. For the detailed proposal on the Road Authority, see: Ibid., pp. 167-176.

29. IBRD, A National Transport System for Nepal, Vol. I (Report and Recommendations) p. 66.

30. HMG, NPC, "The Third Plan (1965-1970)." Kathmandu, 1965, pp. 116-121.

31. Based on interviews with Nepali government officials who were closely familiar with this project. Kathmandu, April-June 1976.

32. IBRD, op. cit., Vol. II, p. 141.

33. R. S. J. B. Rana, An Economic Study of the Area Around the Alignment of the Dhangari-Dandeldhura Road, Nepal, op. cit., p. 21.

34. For the criticism expressed on the choice of the present alignment, and the argument made with regard to the preferability of an alignment from Dhangari to Silghari-Doti, see: Ibid., pp. 24-25.

35. Ibid., p. 29. (See also our discussion on this project in the following chapter).

36. To mention only a few of such local projects see for instance The Rising Nepal (daily newspaper), Kathmandu, March 5, 1976--report on the completion of a four-mile link road between Andhikhola and Jurmi (Gandaki zone). The road was constructed by voluntary labor. March 9, 1976—voluntary work on a 14-mile section of the Phidhim-Ilam road (Mechi zone). March 13, 1976—13 km motorable road linking Damru and Maranghat was built by volunteers. March 25, 1976—a 10-mile road built by voluntary work at Syabruma village (Jhapa). April 17, 1976—10 miles of the 33-mile long Piuthan-Rolpa road, built by voluntary work. The visits conducted in several parts of the country as well as the interviews held with Nepali citizens and officials indicated that it was common for local roads to be built nearby the house of the local strongman.

37. HMG, NPC, "Fourth Plan (1970-1975)." Kathmandu, 1970, pp. 120-122 and 132-134.

38. Ibid., pp. 132-134, and compare with the proposals in the

Master Plan (IBRD, op. cit., Vol. II), p. 116, 124, 127, 131, 135, 138.

39. Ibid., pp. 135-136.

40. UNDP, Road Feasibility Studies, Construction and Maintenance—Nepal: Project Findings and Recommendations. N.Y., 1975 (DP/UN/Nep.-69-516/1), pp. 3-4.

41. See reference No. 24 in the chapter on "Transportation and Development."

42. For proposals and guidelines prepared by the UNDP for the Roads Department, see: UNDP, op. cit., pp. 40-67 and Annexes 5 through 13 (pp. 81-226).

43. Ibid., pp. 15-67.

Chapter 5

PLANNING AND PREPARATION FOR IMPLEMENTATION

Feasibility assessment of a proposed program and operationalization of decisions are part of the planning process. They are preparatory steps which determine, at least in part, the prospects of reasonably efficient implementation. Once completed, these steps usually result in a proposal for the most suitable course of action to implement the program. This includes the specification of operational tasks for each phase of the program, the prescription of adequate solutions to foreseeable difficulties in the execution of the program, projections on required personnel and equipment and the means to mobilize them. It also includes the organization of administrative support as well as finalization of budgetary arrangements. The use of these steps, although applicable to a broad variety of governmental and private undertakings, is particularly important in road-construction projects. The complex technical and administrative operations involved in these projects and the various surveys that have to be carried out prior to actual construction work dictate the necessity to follow these preparatory steps.

In Nepal, deficient planning and inadequate preparations caused considerable difficulties and delays in the execution of the road projects. This situation was mainly caused by the interactions that evolved between the political and administrative factors which comprised Nepal's internal and external arenas. The impact of these interactions on the manner in which the road projects were prepared for implementation is discussed below. The discussion deals first with the internal arena and refers mainly to the Dhangari-Dandeldhura road project. It then proceeds to deal with the external arena and examines the patterns in which the aid-donor countries prepared their respective projects for execution.

The Internal Arena

Among the major road projects only the Dhangari-Dandeldhura road (D-D) was an undertaking in which the Nepali administrative machinery was given substantial responsibility for planning functions and for implementation. Moreover, it was in this particular project that local elites could fully utilize their political weight (limited as it was) to pressure the central regime and the local administration. The D-D road is therefore the only major road project that can serve as an example of the role and impact of internal forces in the planning of an almost all-Nepali undertaking.

This project, which began in 1966/67, was originally planned to be completed by 1973. Construction was still underway in 1980, plagued by numerous administrative and technical difficulties. The present alignment which follows essentially the same alignment proposed by the RTO in 1960, was neither then nor in 1966/67 (when the project was resumed) the subject of a careful feasibility study. In 1960/61 a reconnaissance survey was conducted on the entire alignment, followed by the beginning of actual construction work. In 1966/67, following a decision to revive the project, HMG allocated Rs 1,600,000 for survey and construction work on the alignment. However, during the course of that fiscal year the allocation was reduced to Rs 900,000 of which no funds were spent![1] In 1967/68 HMG allocated Rs 1,800,000 to the project primarily for surveys in the hilly area of the alignment and some construction work in the Terai. The surveys conducted in the hills during that fiscal year were mainly reconnaissance studies and as such they could only be of limited utility for the execution of the project. In 1968/9, when the US resumed participation in this project, it refrained from evaluating this alleged feasibility study.[2] Moreover, during this stage the donor appeared anxious to turn the project into an aid activity that would involve as many local "Panchayats" as possible. Accordingly, the US was not concerned with details pertaining to the future completion of the project.[3] Hence, construction work continued on the basis of the reconnaissance survey.

In 1973, some six years after the project was resumed and amidst numerous implementation difficulties, the road finally became the subject of a careful engineering study. Upon the insistence of USAID, a team of American consultants was invited to conduct the study. In specifying the team's mission it was emphasized that the USAID "particularly desired an independent review of construction status, plans, cost estimates, and HMG's technical capabilities to properly define the project and the work remaining to be accomplished."[4] In June 1973, the team submitted its report and indicated that proper implementation of its recommendation would enable the completion of the project by the end of 1976.[5] However, in April 1976 the Nepali newspapers reported the lack of satisfactory progress in the construction of the D-D road and noted that a new date of completion was set for 1979.[6] In other words, once the long delayed study was completed, it could only solve a few of the many problems with which the project was plagued.

Inadequate preparation for this project was without a doubt a major cause behind some of the delays. There were gross mistakes in cost estimates whereas operation guidelines were too general to be useful in the actual execution of the project. The results of this faulty preparation were pointed out in a CEDA/ILO study conducted in 1973—"the tremendous increasein cost estimate is due to the fact that the original estimates were based, among other things, on the erroneous assumption that the stone crushing for the sub-base and the base could be done along the road. However, recent studies have shown that 1) there are no suitable sites for the location of a crusher plant in the hill section of the road, and 2) that the hard rock required for the base course is not available in any significant quantities along the alignment. Consequently, the base-course material will be crushed in a plant presently being constructed in Km 23 and transported to the site. This necessitates the purchase of 85 additional trucks."[7]

Although both partners in this undertaking should be held accountable for such deficient planning, the bulk of that responsibility falls on HMG. Indeed, considering the manner in which the Nepali authorities performed the planning function in the D-D project, such mistakes were practically inevitable. In making its decision to launch this project, HMG did not have a clear idea about the specific alignment it wished to see constructed. Rather, HMG was only certain of where it did not want the alignment to pass. Second, HMG was more than anxious to see the actual construction work start with no further delay. The pressures applied by the local elite also demanded an immediate start to the construction work. In fact, the local elites were convinced that as long as the regime did not decide on a particular alignment, their own immediate interests could be best served if work started immediately, thus enabling them to "establish facts in the field." With the interests of the central regime and the local population coinciding on the issue of actual construction, other details such as planning and careful preparation for implementation were perceived more as obstacles rather than as necessary preconditions for implementation.

Subsequently, a similar approach was adopted by the bureaucracy. A team of inexperienced officials was sent to the area to collect geographic and technical information. At the project site, the local elite supplied the team with piecemeal, conflicting information that eventually became the basis for the team's final report.[8] The report was inconclusive, skirting practically every aspect of the proposed project, and received only passing attention from the Kathmandu-based officials.[9] These latter had neither the expertise to evaluate the preparations for a large scale road project nor the inclination to examine a project that had already begun.[10]

In sum, this all-Nepali road project was launched with practically no preparatory steps taken to secure its effective execution. In subsequent years, this mischief would entail numerous difficulties in managing construction work in the project.

The External Arena

With the exception of the the US (D-D project), all aid-donor countries considered the stage of preparation for implementation to be an integral part of their responsibility. The donors insisted on carrying it out by themselves with as limited Nepali involvement as possible. There were several reasons for this attitude. First, it was generally understood that Nepal could not perform that complex a task by itself. Secondly, each donor had its own political interests to serve by entering into an agreement with HMG. Each donor probably assumed that by controlling the stages of planning and preparation for implementation of its project it would be able to formulate the most suitable program to protect its own interests This would also allow it to execute the project the way it knew best. Furthermore, in view of the fact that road construction operations usually require adjustment of plans during implementation, each donor wanted control over the preparatory stages in order to retain significant control over the process of adjusting the plans. All aid-donors attained control over the preparatory stage and reduced Nepali participation to a minimum.

With the exception of the Chinese, the planning undertaken by the donors proved most inadequate. Specifically, the factors which determined the adequacy or inadequacy of the respective programs were: (a) the tendency on the part of the donor to superimpose its standard operating procedures on the Nepali system; (b) the extent to which the donor examined circumstances in Nepal in terms of their prospective implications on the execution of the project, and (c) the success of the donor in preparing a program that would enable it to obtain operational independence without antagonizing the recipient.

In preparing a project for implementation, China appeared to be the most effective donor. Shortly after the agreement between the two countries was signed, a team of Chinese technicians conducted a comprehensive study of the area surrounding the proposed alignment (Kathmandu-Kodari). The team surveyed the terrain for a suitable road bed and examined the technical-technological difficulties it might encounter during construction. It studied the availability of local labor and proposed the principles by which labor would be recruited and remunerated. The team collected information pertaining to equipment maintenance and determined what equipment would be needed in the course of the operation as well as when and how supplies would reach the project's site. Prior to implementation, China also considered the recipient's administrative capability and its possible effect on construction work. The Chinese then prepared a program that would not necessitate administrative support from Nepal's weak bureaucracy. Furthermore, in preparing the organizational arrangements, the Chinese made certain that Nepali personnel would be given the official status of equal partners. At the same time, they also ensured that no Nepali official assigned to the project would be able to exert any significant control over any part of the operation without the prior consent of Chinese officials.[11] In subsequent projects in Nepal, China used the same approach and some of the same personnel and equipment employed in its first project.[12]

The approach adopted by the USA in the D-D project lacked China's elaborate preparations. As donor, the USA took almost no preparatory steps to secure a reasonably efficient operation. In fact, the US appeared inclined to ignore numerous foreseeable obstacles in the management of the road project in Nepal. The 1973 feasibility study attested to the fact that the project had been the subject of an engineering analysis some six years after the operation was launched. Second, the study indicated that the donor had intended the operation to be a training ground for Nepali engineers, technicians and administrators but that it failed to design training programs for the participants. Information pertaining to essential aspects of a construction operation—the geo-economic characteristics of the area along the alignment, availability of labor and supplies, equipment specifications—was not collected prior to the actual construction work. Similarly, there were no preparations made for either the recruitment of trainees or the establishment of training systems and facilities. The donor did not review the recipient's capacity to manage the operation and did not engage in a pre-implementation evaluation of the prospects for a successful construction operation. Furthermore, the USA agreed that the recipient's standard operating procedures (SOPs) would govern the project but failed to consider the administrative and financial implications this might have on its own involvement in the project.

In their respective efforts to prepare a road project for implementation, India, the USSR and Britain followed a similar course of planning. Prior to construction, each conducted a preliminary survey of the area along the alignment. Each donor then prepared a detailed program for the construction of the first part of its project. Once construction was begun, each donor proceeded to prepare a detailed program for the next part of the road. Throughout the operation, all three donors insisted that the preparation of cost-estimates, designs and job specifications be handled by officials from the donor country.[13] In preparing the implementation of their respective projects, India, the USSR and

Britain emphasized the technical aspects of the operation but were somewhat less diligent in dealing with issues pertaining to the recipient's administrative capabilities and political sensitivities.

India's performance during the implementation stage would suggest that the feasibility studies it conducted were incomplete. In several cases, for example, low quality materials were supplied in areas known to be vulnerable targets of the monsoon rains.[14] In the case of the eastern part of the East-West Highway, the equipment supplied was totally inadequate.[15]

Britain, in preparing for its project, emphasized the technical aspects of operation. It assigned officials from the British Department of the Environment to carry out the job and insisted that the donor's standard operating procedures should be applied to all aspects of the operation.[16] This approach enabled the donor to maintain control over the preparations for implementation and to reduce the recipient's participation to a minimum. However, by following a predominantly technical approach, the donor overlooked many other aspects involved in the execution of road projects in Nepal. Specifically, Britain did not investigate in depth the availability of labor and appropriate recruitment techniques. Second, the donor failed to consider the relationships between laborers and localpoliticians and the impact that the recruitment of experienced indian contractors and skilled workers might have on the operation. In the course of executing the project, Britain was confronted with the consequences of its "technical" approach. It faced labor unrest that stemmed from unsatisfactory work conditions and had to deal with three strikes that were allegedly incited by local politicians who disapproved of hiring Indians for the project.[17] These labor disputes increased the cost of the project and delayed completion.

Similarly, the USSR applied its own SOPs to the operation. However, the Russians apparently conducted a pre-implementation study on both the technical and non-technical aspects of the proposed project. On the basis of this study, the USSR determined that the location of the project would facilitate the use of capital intensive construction techniques with which it was familiar and over which it could establish control without interference by the recipient. Second, the USSR designated HMG's Roads Department as a general contractor, based on the donor's assumption that the hiring of a local agency familiar with local customs would eliminate complex negotiations with different Nepali agencies.[18] The donor also assumed that it would be able to deal effectively with one local agency and that its dependence upon the services of this agency could be limited. Furthermore, to secure a measure of flexibility, the USSR also reserved the right to award contracts to private entrepreneurs, should it find the Roads Department's services unsatisfactory. These steps might suggest that the Russians considered various aspects of road construction operations in Nepal and prepared a program consistent with local circumstances as well as with the donor's interests. However, given the difficulties encountered during the execution of the project, we find that the Russian program was seriously inadequate. Specifically, the donor failed to arrange the supply of spare machinery parts and found itself dependent on the services of local agencies is more ways than had been expected.

Recent Developments

The substantial extent to which the recipient was excluded from participation in the planning of road projects began to diminish in the 1970s. With somewhat improved administrative capabilities and in view of the experience gained by Nepali authorities, the country began to undertake the construction of some major road projects with its own recourses. Nepali agencies are currently involved in the construction of the Nepalganj-Surkhet road and the reconstruction of the Hetaura-Narayangha road. Although the availability of experienced skilled personnel is still too limited for a fully independent feasibility study to be carried out by the Nepalis, a growing participation of the latter in that process is evident. Moreover, Nepal's request to UNDP to conduct detailed feasibility studies on its secondary road system suggests that HMG increasingly recognizes the importance of preparation to proper implementation. It is still too early to judge the extent to which this awareness is being translated into operational practices, although it is probable that the highly deficient pattern of planning evident in the D-D road will be at least partly avoided. Yet, in view of the fact that the internal factors that influenced Nepali decision-making and performance are similar to those that prevailed during the 1960s and early 1970s, it can also be expected that certain aspects of any given plan will be deficient and that the plans will not necessarily determine the course of the actual operation. The struggle between different administrative units over the distribution of power and resources, the pattern of interplay between the bureaucracy and the palace, and role of local interests are still far too dominant to permit planning to function as a major determinant in the operation.

REFERENCES

1. R. S. J. B. Rana, An Economic Study of the Area Around the Alignment of the Dhangari-Dandeldhura Road, Nepal. CEDA, Kathmandu, Nepal, 1971. p. 21.

2. Based on interviews with Nepali officials who were involved in different capacities in the project (interviews held in Kathmandu during 1976). Additionally, the mere fact that the Americans did not inquire too deeply into the matter at that particular time could not be better explained than by the 1973 study and its findings.

3. Ibid., Kathmandu, Nepal, 1976,

4. "Engineering Evaluation, Western Hills Highway Project." Prepared by Hosking-Western-Soneregger, Inc. Lincoln, Nebraska for USAID, Washington, D.C., June 1973. p. 2.

5. Ibid., pp. 7-59.

6. "Rising Nepal," April 17 and 18, 1976 (A Nepali daily newspaper. Kathmandu, Nepal).

7. H. O. Rieger and B. Bhadra, "Comparative Evaluation of Road Construction Techniques in Nepal." CEDA, Kathmandu. 1973 Part II, pp. 222-223.

8. Interviews held with past and present officials in the Ministry of Transportation. Kathmandu, Nepal. 1976.

9. Ibid., Ibid.

10. E. B. Mihaly, Foreign Aid and Politics in Nepal: A Case Study Oxford University Press London, 1965. p. 155-156.

11. Based on interviews held with Nepali officials who were familiar with the project. Unfortunately Chinese officials at the embassy in Kathmandu refused to be interviewed.

12. Apparently the Chinese continue to employ that principle and it was recently reported that on December 26, 1976, an agreement was signed between China and Nepal for the construction of the Gurkha-Narayangha road. The communique mentioned the fact that "Road construction equipment used in the Ring Road will be shifted to this new project. On completion of this road, work will be taken upon the construction of the Pokhra-Surkhet Highway." Nepal Press Digest, Kathmandu. January 3, 1977 Vol. 21, No. 1, pp. 10-11.

13. H. O. Rieger and B. Bhadra, "Comparative Evaluation . . ." op. cit. p. 210-211 (Britain), 219 (India), 229 (Russia).

14. Ibid., p. 220.

15. Ibid., p. 219.

16. Ibid., pp. 211-212.

17. Ibid., p. 212.

18. Ibid., p. 229.

Chapter 6

THE INTERNAL ARENA AND THE PROCESS OF IMPLEMENTATION

The exclusion of the Nepali administrative machinery from participation in the preparation of implementation programs had far-reaching implications. It created a situation in which the bureaucracy felt little commitment to adhere to plans that failed to protect its interests. Furthermore, it enhanced the tendency among the indigenous participants to consider the stage of implementation as a period of time during which they would compensate for the "losses" they had suffered at earlier stages. This attitude, combined with the actions of the politically-oriented forces within the Nepali bureaucracy, led to the chaotic implementation of transport projects. This involved serious bottlenecks in organizational communication, an increasingly complex and ineffective system of control mechanisms, erratic practices in the management of resources and futile efforts to achieve coordination among administrative units.

Implementation and Organizational Communication

Information pertaining to administrative activities is a commodity as well as a means utilized by various bureaucratic units and organizations to accomplish their tasks and interests. In Nepal, the use of information created serious breakdowns in the communication between administrative units and, subsequently, undermined the execution of projects.

This problem was particularly apparent in the D-D project. As pointed out earlier little was actually known about the circumstances in the area along the alignment prior to the launching of the project. Evidently, little information was volunteered to the center during the implementation stage. This is demonstrated not only by the fact that an engineering study was conducted some six years after the project was launched but by what transpired prior to that study and afterwards. Progress reports, when sent from the project to the center, usually contained inaccurate information on most items pertaining to implementation. No adequate records were maintained on the labor force employed in the project.[1] Difficulties in carrying out the work program were only rarely mentioned, and then only on those occasions when the blame could be placed on delayed supplies of material from India, on weather conditions, or on funds which failed to arrive from the center.

Reference to administrative problems emanating from lack of coordination among units at the center was avoided at all costs. Utilization of equipment on the project was reported only on those few occasions when performance matched projections, whereas records on the maintenance of equipment were either nonexistent or inaccurate.[2]

The lack of proper reporting from the project was caused by deficient organization of the administrative units at the project level, the scarcity of means of communication as well as poor administrative practices on the part of both project personnel and the higher echelons at the center. In the project, the engineer in charge was often overburdened with detailed administrative work. This stemmed from the unwillingness of his subordinates to make decisions and the scarcity of experienced and capable administrative assistants who could prepare reports. Consequently, reports on activities were not only delayed but were also written with little regard to the actual activities in the field. Moreover, the authority given to some of the project personnel to award certain contracts (up to a hundred thousand rupees) led to a situation in which corruption flourished and became a means to compensate for low salaries. This ultimately necessitated the falsification of documents and led to a form of reporting in which the fabrication of information became a common practice.[3]

Furthermore, caution in reporting was essential in order to maintain proper relations with the center, since both the Roads Department and the top echelons of the ministry of transportation were allegedly involved in questionable practices that involved the awarding of large contracts and the procurement and distribution of heavy equipment. The rule of thumb for project personnel in their reports, therefore, was to avoid the risk of stepping on their superiors' toes by omitting any reference to problems caused by decisions made at the center. Indeed, when the center ordered equipment for the project, the project personnel rarely questioned the decision or suggested any changes that would be compatible with the project's needs. For instance, when the reasons for purchasing a crawler crane, concrete post tension equipment or additional scraper were obscure, the project personnel did not inform the center that this equipment was totally unsuitable for the project.[4] Consequently the equipment, purchased at great cost, would lay idle at the project site (a matter not promptly reported to the center). This obviously delayed implementation at inflated project costs.

Exchange of information at the center itself was similarly deficient and the processing of incoming data almost nonexistent, further contributing to difficulties in implementation. J. S.

Gallagher,[5] who reviewed the Planning and Evaluation section of the Transport ministry, wrote, "The staff presently assigned to the unit has neither the professional qualifications nor the interest to do effective work. As a consequence it occupies itself with 'safe' paper shuffling which keeps everyone busy but at the same time avoids the necessity of making decisions or taking action. In such an environment the critical work is inevitably shifted to the Secretary, who is already overloaded with problems from other divisions stemming from the same cause."[6] This practice of paper shuffling implied a lack of substantive data analysis. This in turn entailed the preparation of inaccurate reports that served as background information for decision-making by the top echelons. As a result, their decisions often proved inadequate. However, decisions made by the leaders of the organization "travelled" down the hierarchy and were rarely questioned or disputed irrespective of their content or quality. The lower echelons often found that these decisions were incomprehensible or could not be implemented. Thus, the "memo system" was reactivated.* In many cases, problems were ignored in the hope that they would fade away with time. The impact of these practices on the execution of projects involved increases in the cost of the operation, delays in completion of work and growing management difficulties.

The communication bottlenecks at the center, combined with the tendency to make decisions without considering circumstances in the field, led to additional costly mistakes. This was demonstrated in the D-D project when contractors were awarded contracts although the bids they submitted were almost 50% lower than what the Roads Department considered essential for accomplishing the task at hand. This occurred despite warnings from the Roads Department that it would lead to default by the contractor shortly after work commenced.[7] These predictions materialized and additional delays resulted. In other words, even when exchange of

---

*During interviews held in Kathmandu with several government officials occupying different positions in the hierarchy, the story of the "memo system" was a recurring theme. The term "memo system" stands for the practice in which every official tries to avoid making decisions by simply adding his signature to a memo received from a subordinate and then passing it up the ladder to his superior. Consequently, a short memo is often accompanied by a long list of signatures until it reaches the top echelons who were then forced to make a decision. It then travels back to the lower echelons with minor clarifications added on its way back down the ladder, but with another list of signatures attached.

information did take place the network was such that decisions frequently rested with less qualified units.[8]

The same factors that impeded communication within the system also hampered the process of project evaluation. The flow of inaccurate information through the system, a lack of substantive feedback, and a tendency on part of bureaucrats to submit noncommittal reports made it highly improbable that the evaluations would be an effective means of correcting deficiencies. If fact, evaluations made at the project usually amounted to figures on percentages of work accomplished, with no target specified| It thus remained unclear to all concerned whether progress was above, below, or on target.

This also created a situation wherein the center was unable to take the necessary measures to improve performance. Furthermore, when similar practices were employed at the center, the evaluation of activities became more of a ritual act than an effective means to detect faults in the system and correct them. The data received at the center was rarely examined in the context of means and needs for improved efficiency. Rather, it was usually translated into charts that were presented in a "Review Room" at the departmental or ministerial level, with no understanding as to how accurate they were. Charts on the number of workers employed at a given project, the extent to which equipment was utilized, how much work had actually been accomplished, the cost of maintenance and the impact of delays were frequently unrelated to one another (if and when they were available at all). The process of evaluation therefore amounted to little more than the drawing of colorful charts.

However, this is not to suggest that most participants did not know at least part of the true picture. In fact, most problems were recognized, either in detail or in general, but almost no one was willing to take action. The prevailing rule was "keep the boat afloat or we will all drown." A project manager would not point out the deficiencies of the Roads Department's contract award system, and the Roads Department would not question a decision made by the Secretary. Each participant sought a share in the distribution of resources and each was clearly aware of where his "sphere of influence" began and ended. A critical evaluation could disturb the status quo and no subordinate would seriously entertain the idea of becoming engaged in such criticism, especially in view of the fact that the government was the major employer in the country. Given this background, it can be easily understood why critical evaluations of the D-D project were conducted by two agencies unrelated to the project, the department, or the ministry. In 1970/71, CEDA* conducted an economic study of the area around the alignment, critically examining development projects in this part of the country and referring to the limited contribution of this road project to developmental needs.[9] In 1973, an American consulting team conducted an evaluation of the project from an engineering perspective.[10] In both cases, the lack of data and the "erratic efficiency, organization and cooperation of the agencies collecting or storing various types of data" were pointed out as a constraint upon evaluation and proper execution of the project.[11] Of significance was the fact that almost no reference was made to the political-administrative considerations involved in the exchange of information, although it was quite apparent that the researchers were aware of the magnitude of the problem.[12]

Problems in communication were not restricted to the D-D project and could be found in most other Nepali undertakings in the transport sector. These included the railway services, the ropeway, and some bus services that fell under the jurisdiction of the National Transport Organization (also known as Nepal Transport Corporation or NTC).[13] Commenting on communication in the NTC, CEDA reported:

> There is a total lack of communication in the organization. Weekly staff meetings are held with Division Chiefs and officers. There is, however, no agenda nor minutes of these meetings which have become a weekly, aimless ritual. The Executive Chairman rarely holds joint discussions with the Division Chiefs on the affairs of the corporation. [Rather], he meets with each of them individually. [Consequently], no division chief seems to be able to comprehend the total perspective of this corporation.

The report further added that "Inter-division communication is also complicated through the writing of formal letters and memos, [whereas] informal, conversational problem solving or information dissemination is unheard of in the NTC."[14] J. S. Gallagher, referring to one of the NTC's Division which incorporates the Nepal Railway and the trucking operations (Narayani Division) wrote "The Division is urgently in need of attention because of poor management over a long period of time coupled with widespread malpractices. Both

---

*The Center for Economic Development and Administration (CEDA) at Tribhuvan University, Kathmandu. It should also be pointed out that several of the Center's top personnel have been called upon to serve on the JBK.

the railway and the trucking operation are suffering the consequences of unskilled and inept management. Reliable financial records do not exist and my efforts at building them over the past two years have been repeatedly frustrated."[15]

The causes for deficient communication practices appeared to be similar to those that plagued the D-D project—lack of skilled personnel, unwillingness to take any action that would require the writing of critical reports and widespread administrative malpractices that entailed the fabrication of information. In the case of the NTC's operations there were also allegedly unlawful transactions on the part of Indian trucking companies that controlled a large part of the trucking industry in Nepal.[16] It should also be pointed out that, once again, a unit outside the organization was called upon to perform the task of evaluation (instead of a unit within the administration).[17]

Recent Developments

In view of the unsatisfactory progress in most Nepali projects, the palace, the NPC and the Ministry of Finance attempted to employ different measures that would alleviate some of the communication bottlenecks However, they also had to consider the potentially excessive political costs of some of these measures. This precaution ironically created breakdowns in communications and interfered with the full implementation of corrective measures.

The palace received reports and evaluations from the bureaucracy, but recognized that these data were far from complete and often inaccurate. However, the palace has always maintained its independent sources of information.[18] The establishment of the JBK and its actions as a data-gathering agency, provided the palace with an effective means to obtain information. On those occasions when King Birendra was about to visit a certain region of the country for "review and inspection of development activities," for example, the JBK sent people into the field to qualify the reports which the palace had received from the concerned ministries on their respective activities in the region. They examined the actual circumstances at the projects as compared with these reports.[19] The King was then briefed so that he could show, upon arrival at the project site, that "big-brother" was watching.[20]

The need to employ such practices obviously reflected the dissatisfaction of the palace with the data received from the ministries. In fact, it could be considered a signal to the ministries and project managers that their reports should be better prepared and based on more accurate information, since the palace was better informed than they suspected it to be. Frequently, however, this only further exacerbated difficulties at the project and the ministerial levels, causing ministerial and project personnel to be even less willing to make decisions on activities pertaining to their respective operations. What's more, it could cause an overflow of information wherein every small detail of the process was reported, thus clogging all channels. Considering the limited capabilities of the Nepali administration to process data, any substantial flow of information could lead to a serious slowdown. Pressing the bureaucracy too heavily could result in increased dissatisfaction among large number of bureaucrats whose loyalty and support the King needed. In other words, in a system where a large part of the distribution of power and resources takes place within the bureaucracy, the monarch must allow the bureaucrats certain discretionary power if he is to retain control over the system. Although some deficiencies in communication could be alleviated to enable more efficient implementation, other communication bottlenecks had to be left intact to ensure the political survival of different internal actors.

The question of efficiency versus a share in the distribution of power as it pertains to communication among internal actors also surfaced in two proposals made by the NPC and by the Ministry of Finance respectively. Although that both units initiated their proposals primarily as part of an effort to overcome deficiencies in communication, the proposed measures were interpreted quite differently by the ministries. In the case of the NPC, a proposal was made to conduct a "pilot-test" on different procedures for project evaluation. The NPC suggested that it would send some of its people to a dozen projects in different sectors to conduct in-depth evaluations. Furthermore, it was proposed that the NPC personnel would work together with project personnel to prepare progress reports. This proposal met with substantial opposition from the ministries that considered it an outside attempt to control their operations. Evidently the proposal never became operational.[21]

In the case of the Finance Ministry a proposal was made to send officials (from either the ministry itself or from the office of the Auditor General) to the projects to follow the various stages of implementation and, together with project personnel, to manage the financial records of the operation. Again, the stated objective was to improve the financial administration of projects and to establish proper financial reporting procedures as well as reliable financial records. But the ministries, already alarmed by the increased power of the Finance Ministry, felt that this was another step to tighten control over their operations. They opposed the proposal and it was shelved.[22] The question of how the ministries would have reacted had these proposals been implemented seems to have been answered by the manner in which the order to establish planning cells was executed. In sum, the data pertaining to activities in any given project or ministry and the manner in which it was transmitted to other agencies constituted an important component of the power these units possessed. Any attempt to change this part of the administrative process was perceived as a threat and encountered opposition. Increased efficiency in the execution of projects or improvements in reporting and evaluation practices were not effective arguments to convince the ministries to change their mode of operation.

Implementation and Administrative Control Mechanisms

Difficulties in establishing effective control mechanisms within the system emanated from the same factors that led to inadequate communication among internal actors. In addition to limited administrative capabilities, there was the tendency among internal actors to perceive administrative acts mainly in the context of their immediate or expected effects on the distribution of power and resources within the system. This led to a "vicious-circle" in which the introduction of one set of control mechanisms was almost immediately met by counter measures taken by those actors whose activities might otherwise have been affected. It subsequently stimulated additional efforts to improve administrative control within the system, resulting in an even stronger conviction on part of the potentially affected actors that the main purpose of the exercise was to dispossess them from some or all of their share in the distribution of power and resources.

The issue was mainly one of financial control over operations, i.e. in Nepal, financial arrangements were traditionally determined by control considerations.[23] The complexity of the process of releasing funds to projects (after they were authorized), the multiplicity of accounting systems (excluding those of the aid-donors), and the rigidity of regulations and requirements pertaining to these processes made smooth execution of projects almost impossible. When accompanied by limited administrative capabilities, these regulations almost inevitably led to frequent delays in operations. Moreover, in some cases these delays led to the necessity to start a project anew several times. The ministries, in an effort to maintain a degree of independence and to pursue their operations, adopted a rather simple technique. They deposited the money which was not spent during a certain period of time in bank accounts, to be used when they deemed it appropriate. When the Finance Ministry discovered the scope of this practice, it closed all accounts and insisted that every year begin anew financially in order to flush out funds which the departments were holding but not using. The ministry also insisted that before funds were released for the third quarter, the departments had to submit evidence of having used up the first quarter's allotment. For their part, the departments found they could neither keep up with accounting nor with spending. Since accounts in Nepal are based on monies actually disbursed, with the emphasis on meticulous accounting of funds spent and funds on hand, the process proved enormously time-consuming. As a result of cumbersome accounting techniques, the departments usually handed in their first quarter statements late. Moreover, these statements often failed to satisfy the Finance Ministry. This meant that some of the work had to be done over again and the release of funds was further delayed. The "vicious circle" of underspending therefore continued since the departments found it difficult to spend these funds by the end of the fiscal year.[24] The impact of these delays on the execution of projects was considerable. The projects which were affected the most were those that involved construction work. The unavailability of necessary funds prevented the completion of construction work before for the rainy season. Since the monsoon rains often caused land-slides, some work that had already been completed would have to be done again. This was indeed the case on the D-D road as well as on the Russian project.[25]

The emphasis on control and accountability combined with the lack of competent personnel at the middle level management and in the lower echelons, produced peculiar complexities in the governmental process. A unit chief in the NTC was authorized to place orders for equipment in an amount not to exceed Rs 2,000 at one time. To order 5 truck tires which cost a total of Rs 10,000 (Rs 2,000 per tire) an official had to place 5 separate orders. Although detailed and demanding, "The financial control system of NTC is very poor. The accounting system is not complete. It takes a long time to gather information from different units. The profit and loss account and Balance Sheet of the Corporation have not been prepared for the last two fiscal years. NTC has no control over revenues, debt collection and costs of operation mainly because the accounting procedure is incomplete . . . [Similarly, since accounts were not up to date, financial planning was practically nonexistent.]"[26]

These issues became even more complicated when a project involved external grants and loans or, worse yet, when it involved Nepali financial participation in the implementation of the project. In cases involving Nepali financial participation (represented by the D-D project), as well as those undertakings which donor depended on some Nepali technical or administrative support (like the Russian project and to a smaller extent the British project), the operational and accounting techniques of the donors became part of the picture.

This not only increased confusion among Nepali administrators but on those frequent occasions when the donors' accounts of unspent balances and the departments' accounts were incongruent, additional time was consumed to reconcile the disparate accounts. Furthermore, with the donor's fiscal year differing from the Nepali fiscal calendar, a situation evolved in which one of the two partners (usually Nepal) could not supply updated information on activities and expenditures on a timely basis. This led to additional misunderstandings between the partners and inherent constraints in the preparation of work programs for the following year (uncertainty as to whether funds that had lapsed would be available for ongoing activities, the need to revise cost estimates due to inaccuracy of data, assessments of the effects of delays on the project's total cost). The intricate web of control procedures and practices was further complicated by the precautions which the Ministry of Finance and the donors took to secure proper utilization of external funds.[27]

The broad variety of control mechanisms were only of limited effectiveness in the elimination of poor administrative practices. The fact that the country's administrative machinery was highly centralized and politicized, a priori undermined the prospect that control mechanisms could be perceived in purely technical-administrative terms. The limited administrative capacities of the bureaucracy, presumably necessitated an emphasis on control, but constituted a major impediment to efficient application of these mechanisms. Hence, the measures adopted not only failed to contribute to improved execution of projects but often caused an increased waste of scarce resources.

Implementation and Coordinated Effort

The unsatisfactory coordination between Nepali agencies was caused, in part, by faulty planning.[28] However, a factor of much greater impact was the competition between agencies seeking to either gain or maintain control over the distribution of resources. In road projects this involved such issues as jurisdiction over procurement and distribution of supplies and equipment, the authority to issue construction contracts, and control over the assignment of work to various units and sub-divisions. Control over these activities constituted power within the system and an ability to secure a share in the distribution of rewards. Under these circumstances, coordination between participants was usually based on a <u>temporary</u> and implicit understanding between competitors on the boundaries of jurisdiction each enjoyed. This implied excessive flexibility in coordination practices as well as frequent changes in both formal and informal arrangements. This pattern evidently had little to do with coordinated efforts aimed at the smooth execution of projects.

The division of labor between the units involved in the execution of the D-D project was as follows: The project manager was authorized to award construction contract not exceeding one hundred thousand rupees; the Roads Department was responsible for larger contracts, while the top echelons of the Transport Ministry usually took part in decisions pertaining to projects that involved particularly large investments. A similar division of responsibilities existed among these units with regard to equipment. However, in the early 1970s, amidst consistent rumors of flourishing corruption and with implementation lagging far behind schedule, a new administrative arrangement was introduced to deal with the procurement of supplies and equipment. The National Trading Corporation (more commonly known as National Trading Limited or NTL) was placed in charge of procurement and distribution of equipment for all Nepali projects. The responsibility for the purchase of petrol for all projects undertaken by HMG was given to the Nepal Oil Company. These changes, part of an effort to curb the alleged corruption, also enabled the palace to be somewhat more involved since presumably it could more easily control public enterprises located outside the bureaucracy.[29] Still, the manner in which the change in responsibilities was introduced indicated the insufficient attention paid to the operational aspects of the new arrangements, especially coordination between the units involved.

From the outset, it was quite apparent the the NTL had neither the personnel nor the expertise to determine which equipment was actually needed and what priorities should be given to the order of purchase. Hence, the organizational change added another unit to the implementation network without providing the means for this unit to perform its responsibilities. The Roads Department and the project managers, although required to work through the NTL instead of dealing directly with import/export companies, continued to maintain de facto power over purchases—i.e. the alleged corruption could continue, although in a somewhat more complicated manner.[30]

The problems emanating from deficient coordination were even more serious in the case of the Nepal Oil Company which had the responsibility of issuing purchase orders, following them through in India (where all the oil was purchased), and organizing the distribution of petrol to different projects in different parts of the country. However, during the period immediately following the reassignment of responsibilities, a situation emerged in which purchase orders issued earlier by the Roads Department and negotiated with Indian companies were not filled despite the fact that a down-payment on the transaction had already been made. Upon inquiries into the matter the Roads Department was informed by the Indian company that action on the purchase order would be taken only through the Nepal Oil Company and that the matter should be cleared with the latter.[31] Although delay could be expected because the Nepal Oil Company lacked experience in these matters, the very fact that the Roads Department and the Nepal Oil Company failed to inform one another of their respective actions indicated that coordination between them was at a minimum. Moreover, failure to fulfill certain purchase orders led to a slowdown in construction activities on the D-D project and delayed completion by several months. In other Nepali undertakings in the transport sector, problems caused by inadequate coordination were also apparent. In the case of the ropeway, high level officials in different ministries occasionally issued orders for priority transport of certain goods, upsetting the entire operation schedule of the ropeway. Moreover, some government corporations, after ordering goods they presumably needed, left the merchandise at the terminal for long periods of time. Thus, not only was the merchandise improperly stored and unused, it blocked the terminal facilities, sometimes halting the operation of the ropeway altogether.[32]

Total disregard of the need for proper coordination was a characteristic feature of the mode of operation of most bureaucrats and administrative units. It was an attitude or approach consistent with a lack of information on the part of the officials about the scope of operations of their own units, let alone other agencies. It was also an outcome of a highly centralized system in which decision-making was avoided by lower echelons while the top echelons who made the decisions often lacked minimal familiarity with the nature of operations they ordered. It was a result of a system in which rewards for efficiency were rarely bestowed upon an official, and the risks of taking initiatives and thus stepping on somebody's toes were too great, carrying with them punishment. Moreover, considering the distribution of power within the system, coordination between different units was either a matter to be avoided or, alternatively, to be so arranged that no one unit would interfere with the activities of the other, allowing both to enjoy the advantages of "cooperation." Efficient execution of projects or smooth operation of a certain service had little to do with those arrangements, unless it became so deficient as to attract the attention of the palace or the Ministry of Finance.

Procurement of some of the equipment was not necessarily done according to the needs of the project, or, for that matter, with an effort to consider the utility of the equipment in similar undertakings or the potential of using it in other operations in the area around the project site. This might have involved further administrative effort to get the necessary approval and the potential of having to share the rewards of a fruitful accomplishment. On the other hand, when coordinated efforts could not be avoided they occasionally assumed the form of keeping the interests of the partners satisfied. If, for instance, two agencies found it necessary to cooperate in the procurement of equipment, with one being at the purchasing end and the other at the ordering end, then, once the goods were purchased and delivered from across the Indian border (e.g. to a dock at one of the terminals of the ropeway), the need for the goods to reach the project site appeared somewhat less urgent. Once the first stages of cooperation had yielded results, both sides allowed other administrative constraints to take their course—i.e. difficulties in obtaining speedy approval for trucks to transfer the goods or delays in receiving the necessary documents for release of the goods. In other words, efficient execution of a project occasionally came secondary to the attainment of a share in the distribution of resources.

Implementation and Personnel

Throughout the discussion thus far, the lack of skilled personnel was mentioned as a serious constraint in practically every aspect of the implementation process. Two major components were pointed out in that context: first, the shortage of adequately trained and competent officials, particularly at the medium and lower levels of the bureaucracy; second, the apparent fact that the interests of most officials were often incongruent with efficient performance of their formal responsibilities. A third major component directly related to that issue was the manner in which the available personnel was organized and utilized within the bureaucracy. The personnel policies and practices of HMG proved less than conducive to efficient performance in the execution of both regular administrative operations and development projects. This involved problems of personnel recruitment, lack of training, improper job assignment, lack of motivation, unwillingness on the part of most officials to make decisions on issues associated with their formal responsibilities, lack of initiative, noncommittal attitude toward work, and an almost nonexistent system of rewards for successful performance. Furthermore, considering the fact that most studies conducted on the Nepali bureaucracy during the last two decades pointed out the inadequacy of HMG's personnel policies and practices and identified those in use as major constraints in the administrative operations, it becomes apparent that little was actually accomplished to change that state of affairs.[33] While evident in every sector of the economy, the ramifications of this situation were particularly significant in the transport sector where the operations required, by their very nature, both technical and administrative skills. Lack of skilled personnel, the inadequate organization of available manpower, insufficient administrative support, and the practices employed by different officials in the course of performing their formal duties led to deficiencies in the implementation of most Nepali undertakings in the transport sector.

The D-D Road Project

In Nepal, as in many other countries, an assignment to a project located far from the center was usually considered a demotion. Indeed, with the possible exception of the engineer in charge and a few other officials, the personnel in the D-D project consisted of inexperienced officials. These were officials who lacked either the ascriptive attributes that could assure them a position in the center or, worse yet, the necessary skills to perform similar duties in presumably more important undertakings. Skilled personnel on the other hand refrained from voluntarily taking upon themselves an assignment in this project. Furthermore, when assigned to that undertaking, officials usually tried to dodge it or concerned themselves mostly with finding a way to return to the center as quickly as possible. The team eventually gathered for this project was then expected to implement plans that were poorly prepared in the first place. Due to the lack of sufficient administrative assistance, the engineer in charge soon found himself attending to a large number of minute administrative responsibilities—writing reports, making decisions on small details of the operation, or dealing with endless memos from subordinates who were unwilling to make decisions themselves. The project manager, who lacked sufficient administrative training as well as the willingness to perform these bureaucratic tasks, soon resorted to another "technique" in coping with this matter. He ignored part of the administrative work and transmitted some matters to units in the center for them to make the decisions. Since many officials in the center were also dodging their responsibilities and unwilling to make decisions, various problems pertaining to the project remained unattended. It took approximately two years before a new manager was appointed to the project and additional aides assigned to the administrative section of the project. Aware of the fate of his predecessor and assisted by a larger number of administrative personnel, the newly appointed manager actually paid more attention to paperwork than to the execution of the project. This did not produce more efficient performance in the implementation of the project; rather, it merely accounted for a larger volume of documents and reports, presumably indicating the propriety of actions taken and disbursements made in the course of the project execution.[34]

The D-D project was portrayed, among other things, as an important training ground for Nepali engineers, technicians and administrators. In fact, inefficiency in executing the project was occasionally justified by this aspect of the operation. The American team of consultants commented that "training could be one of the most valuable and lasting benefits of the project . . . . However, care must be exercised to insure that proper procedures and techniques are being presented, both in construction methods and in the proper utilization and operation of major items of equipment."[35] This comment, made some six years after the project was launched, was a clear indication that personnel training was not actually incorporated in the project plan and that the extent to which any action was taken on the matter was limited to a poorly supervised "trial and error" method of operation in which the training element was at best incidental.

A similar pattern of performance was evident in other Nepali undertakings in the transport sector. In its study on the NTC, CEDA reported that "like in most corporations, the Board of Directors of the NTC, is not alert. The Executive Chairman, in spite of his sporadic endeavors to bring some reforms in the Corporation does not identify his goals with that of the Corporation.[36] The report further pointed out that the then officiating Executive Chairman and two of his predecessors lacked minimal expertise in the field of transportation and their requests to HMG to provide them with training in the management of transport operations had been turned down.[37] Commenting on the corporation's employees the report pointed that in its 10 years of existence, NTC sent only two of its staff members for training. Moreover, it was also found out that in the central organization, 27 out of 82 employees had been on temporary appointments for as long as 7 years.[38] Referring to yet another unit in the organization, it was mentioned that it took almost two years to get approval for the appointment of the corporation's auditors and that the planning section was "manned by people who would find it difficult to fit themselves anywhere else."[39] Personnel problems could also be found in the central bureaucracy. J. S. Gallagher, in his report, argued that it was not only the serious shortage of trained, experienced and interested transport service managers and administrators that constituted a bottleneck in the operations, but rather that "this shortage was aggravated by HMG personnel practices which shift civil servants from one position to another without regard for their training, competence or interest in the new assignment."[40]

The inadequate policies and deficient practices in matters pertaining to the management of manpower resources emanated from the considerable politicization of the administrative machinery. Appointments to administrative positions were usually based on political considerations. The removal of higher echelons after several years (3 to 5) in one position and their reappointment to a different position in the government was largely based on an approach that sought to prevent any single actor from gaining substantial power in one area or field of governmental activity. In addition, new leaders in any organization often brought with them a coterie of proteges to fill different positions. The lower level echelons were well aware of the fact that these appointments were determined according to ascriptive criteria. They also recognized the "rotation" system to be part of the competition for power. Hence, upon considering the prospects for survival in the system, the majority of officials chose to refrain from action and responsibility. The rule of thumb was that the fewer decisions and initiatives an official takes, the higher the probability that he will not be reprimanded for interfering in a superior's realm. The impact of this approach on the execution of projects was considerable since the officials not only refrained from taking action but also avoided making proposals that could improve the operation. A sub-division chief on the Janakpur Railway, for instance, declined to order a reduction of freight charges for return trips from Janakpur to Birganj even though it would solve the problem of empty return trips. Similarly, the engineer in charge of a section of a road project would hesitate to propose the construction of the alignment closer to the bank of the river even though this would improve the alignment.[41]

Added to the impact of these practices were the adverse effects of long and complex formal procedural arrangements pertaining to appointments. It often took from one to two years to get the necessary approval for new appointments and just as long to get a job abolished. Moreover, when authorized, it often happened that new appointees lacked the qualifications required for their assignment. Consequently, many units had to employ personnel on the basis of temporary appointments. Additionally, many offices suffered from inflated numbers of officials on their staff simply because it was too complicated to fire them or to reassign them to other units. Moreover, with unqualified personnel filling various positions in the bureaucracy, qualified personnel was either unemployed or under-utilized. It thus happened that in the ministry of Transport, several engineers were assigned to the administrative section of the organization while some found themselves appointed to other ministries (Housing, Commerce) where their professional skills could not be utilized. The impact of such personnel practices on the willingness and ability of officials to perform their duties efficiently was adverse. It created the inevitable necessity on the part of many officials to resort to a "trial and error" mode of operation. It also furthered an attitude of indifference toward work among qualified officials who found themselves in positions incongruent with their skills. Underemployment within the units affected morale among all officials and workers since it fostered idleness.

The lack at all levels of personnel policies and practices geared to active promotion of initiative, assumption of responsibility and productivity was a dominant feature of the Nepali administrative machinery. Faulty organization, cumbersome administrative procedures and the predominance of political considerations in personnel policies and practices evidently led to serious deficiencies in the process of implementation of development programs.[42]

Other Components and Constraints in the Process of Implementation

Three other facets integral to the implementation of transport projects in Nepal were the maintenance operations in road construction and transport services, the labor force, and the geographical setting and climate conditions of the country.

Equipment and Road Maintenance

Effective road construction operations and the serviceability of established road networks depend, among other things, on adequate maintenance of construction equipment and an efficient road-maintenance program. In Nepal neither of these were accomplished satisfactorily. The Ministry of Transport and the Roads Department failed to establish a strong mechanical section to maintain construction equipment. The procurement policies of these organizations complicated and prolonged the process of obtaining the necessary equipment and spare parts. Improper handling of equipment—e.g. shipment to the project site—reduced its expected productivity. Inefficient inventory management hampered logistics operations. The scarcity of skilled operators increased the frequency of breakdowns and the shortage of skilled mechanics kept the quality and regularity of maintenance at a very low level. Under these circumstances delays in the execution of programs became practically inevitable. In the D-D project, for instance, deficiencies in equipment maintenance were identified as a cause of unsatisfactory progress.[43] Moreover, in 1975 the UNDP's technical assistance team to Nepal indicated that without sweeping changes in the structure and activities of the Roads Department, deficient equipment maintenance operations would increasingly hamper implementation. The team referred in particular to the need to change relations between the Department and the ministry and to the adverse effects that HMG's personnel policies and practices as well as the highly centralized pattern of decision-making had on the execution of projects.[44]

A similar state of affairs existed in road maintenance programs. The planning of maintenance operations was minimal and funds for these activities were inadequate. Studies conducted by several external agencies identified these deficiencies as emanating from HMG's lack of interest in this aspect of road construction programs.[45] In the early 1970s, amidst growing difficulties to keep roads in service, HMG requested the assistance of the World Bank. The four-year maintenance program that was subsequently launched proved insufficient. In 1975 a report prepared by another external agency indicated that due to the lack of adequate maintenance, many roads were in danger of becoming unusable. The report identified managerial shortcomings and cumbersome administrative procedures as impediments to efficient planning and execution of the maintenance operations.[46]

Labor Force

In addition to political and administrative constraints, difficulties in establishing sufficiently large and skilled labor forces for road construction operation also hampered the execution of projects. Most of the Nepali labor force employed in road construction was described as undependable and inefficient and HMG was less than earnestly involved in any effort to establish corps of skilled workers in this field.[47] This was a consequence of the economic structure and the relative advantage that Indian contractors and experienced laborers had over their Nepali counterparts. To most Nepali workers, employment in road construction provided a supplementary income to their earnings from agriculture. Consequently, they did not seek to gain expertise in construction operations, nor did they wish to secure permanent employment in such projects. Quite frequently, in fact, once a Nepali worker earned enough money for his needs, he left the project to return to his village. Moreover, if in any given year the agricultural harvest proved adequate, the worker usually refrained from seeking additional employment. In both cases the actual work program and the availability of the labor force were problematic, often leading to delays in operations and difficulties in meeting time schedules. It also created a tendency among aid donors to seek the services of Indian contractors and laborers who usually were more experienced and accepted such employment as their main occupation. Indeed, with the exception of the Chinese projects and to a certain degree the D-D project, Indian laborers were employed on other projects, particularly for tasks involving strenuous work (lifting, hauling and manual stone crushing) or professional skills (bridge building, construction of retaining walls and culverts). Lack of local laborers or unwillingness on their part to partake in this work necessitated the recruitment of labor from other districts, sometimes as far as 200-300 miles from the project site.[48] This occasionally added costs to the execution of the project and aggravated the difficulty of obtaining sufficient of laborers.

The relatively large number of Indian workers and the apparent dominance of Indian contractors in the construction industry affected the execution of projects in more than one way. It was in a sense an answer to the lack of indigenous construction capacities, but Nepali animosity toward Indian dominance, combined with some questionable arrangements pertaining to the awarding of contracts, had an effect on workmanship and cost in that industry. Labor unrest and strikes were caused by Indian participation in certain projects.[49] The manner in which construction contracts were awarded led to a situation in which financial backing became a more important criterion than technical expertise, which subsequently led to faulty performance by contractors and allegations of corruption among the officials in charge.[50] It was against this background that a 1975 UNDP report called upon HMG to set the establishment of an efficient contracting system as a development objective. It recommended that the government pursue an active policy of fostering the infant industry and set up a control system that would avoid monopoly situations and prevent the formation of contracting "rings."[51] Though the number of Nepali contractors has increased in recent years and public corporations like the National Construction Company of Nepal (NCCN) began receiving more contracts, the issue remains far from resolved and most of the UNDP's recommendations are yet to be translated into operational terms.

Geographical Setting and Climatic Conditions

Road construction operations in Nepal were further complicated by the country's geographical setting and its climatic conditions. The rainy season which lasts from June to September usually prevents construction work during that time of the year. Additionally, since the peak agriculture season often extends into the month of October, labor force was usually available only 7 months of the year. Furthermore, the monsoon rains occasionally caused serious landslides on roads under construction, particularly in the hills, and often slowed down work on other sections of the road. Lack of adequate transport facilities and the remote and inaccessible location of some projects made the transfer of equipment and supplies to the projects' sites a complex operation in itself, often causing additional delays. Technical and technological difficulties were encountered in construction works in the hills, as Nepal's major river systems (Karnali, Gandak, and the Kosi and their numerous tributaries) made the construction of major bridges (Kamala, Narayani) and minor bridges necessary in every road construction operation, adding technical and logistical difficulties to most of those undertakings. In brief, though usually not obstacles of substantial magnitude, the geographic setting and climate conditions added to the road construction problem in Nepal.

To sum up, the limited administrative capacity to undertake large scale projects and the highly politicized bureaucracy undermined project implementation in Nepal. It involved practically every element in the internal arena and affected all components of the implementation process. The execution of road projects was part of the competition among internal actors over the distribution of power and resources. Therefore, most efforts to achieve efficiency in the implementation process were thwarted.

REFERENCES

1. Based on interviews with officials who were employed in the project in different capacities. Kathmandu, Nepal, 1976.

2. On the lack of maintenance records see "Engineering Evaluation, Western Hills Highway Project" USAID, op. cit. 1973 p. 33.

3. A rather "safe" item about which information could be fabricated, gaining additional income, was the hiring of workers. Some of the project personnel, either by themselves or in collaboration with contractors, reported inflated numbers of workers to whom salaries were presumably paid and pocketed the difference between reported payments and actual payments. This was done in the following way: The workers, most of whom were illiterate, were requested to certify receipt of salary. Unaware of what they were signing, they evidently certified (by finger print) double or triple receipts of salaries. These documents constituted the records of payment which were transferred to the accounting offices. Needless to say, the workers received their salaries only once, whereas the administrators kept the rest of the money for themselves.

4. On the the procurement of unnecessary and inadequate equipment see, Ibid., pp. 35-37 and, H. O. Rieger and B. Bhadra, "Comparative Evaluation . . ." op. cit., 1973. p. 252.

5. A transport economist who as a UN expert was assigned in the early 1970s to assist the Ministry of Transport, HMG on various transport matters.

6. J. S. Gallagher, Jr. "Eighth Report to HMG and to UN Office of Technical Cooperation" (for the period July-December 1972.) Kathmandu. January 1973 p. 2.

7. H. O. Rieger and B. Bhadra, "Comparative Evaluation" op. cit. p. 224.

8. In the interviews held in Kathmandu with various well-informed sources the recurrent theme mentioned by almost all people was that the leadership of the ministry itself was occasionally involved in questionable practices pertaining to distribution of contracts particularly when the sums involved were large. Such allegations could not of course be verified considering the very nature of such alleged practices. It should, however, be noted that the rules of the ministry do not <u>ordinarily</u> permit the award of contracts to any but the cheapest tender.

9. See, R. S. J. B. Rana, An Economic Study of the Arena around the Alignment of the D-D Road, op. cit. 1971.

10. "Engineering Evaluation, Western Hills Highway Project," op. cit. 1973.

11. R. S. J. B. Rana, An Economic Study . . ., op. cit. p. 27 and, throughout the "Engineering Evaluation," Ibid.

12. Based on interviews with Nepalis and Americans who were involved in those studies.

13. The National Transport Organization was established upon approval of the National Transport Organization Act in August 1965. In the provisions of the Act, it is explained that the Organization was created in order to "make arrangements to properly organize and coordinate the management of the transport services in the country, encourage the development of such services in a systematic manner and provide the maximum possible transport services to the general public so as to ensure their comfort". For an account on the organization and functions of the NTC, see Sir Eric Franklin, "Special Report on the National Transport Organization." UNDP, XR/UNTA/17. Kathmandu, (N.Y.), August 1966.

14. "Study of the Transport Corporation of Nepal," CEDA, Tribhuvan University, Kathmandu, 1973. p. 3.

15. J. S. Gallagher, "Eighth Report . . ." op. cit. 1973, p. 6.

16. In interviews held with Nepali officials as well as with foreign experts familiar with the subject matter it was frequently pointed out that one of the reasons for the ropeway operating at only 20-30 percent of its capacity was that there were "incentives" provided by the Indian trucking companies to those in decision-making positions to keep the ropeway under-utilized (see also section in "Implementation and Coordination").

17. The regime apparently found it most suitable to use a unit like CEDA for purposes of conducting evaluations on different governmental agencies and undertakings. CEDA had highly skilled and well-educated personnel who evidently could undertake such studies. Furthermore, considering the fact that the members of the staff belonged to academic circles rather than to the bureaucracy, their work was presumably that of impartial observers.

18. Little is actually known about the inner workings of the palace,

but it is safe to assume that the palace was generally better informed than many bureaucrats might have thought. It appears as though the palace maintained sources within the bureaucracy, as well as among some of the local elite, the details of which could not be verified. Additionally, it appears that King Mahendra brought the "art" of divide-and-conquer to a rather high level of efficiency both within the palace and outside it, and was able to obtain impressive amounts of information on a variety of operations. Nevertheless, even he could not overcome many communication bottlenecks. Of course there were occasions on which he chose not to act, for reasons of maintaining a certain status quo within the system or for the purpose of allowing newcomers to balance the power of others within the system.

19. Based on interviews with various well-informed sources in Nepal (Kathmandu, 1976). It should also be noted that King Birendra took an active role in the matter by visiting several areas each year and reviewing the development activities undertaken at each location. For instance, in January-February 1976 the King visited the Far Western Region, staying there for almost a month (particularly in Surkhet,) whereas in April he visited the Central Development region reviewing various projects in the Bagmati, Narayani and Janakpur zones.

20. Based on interviews with different government officials (Kathmandu, 1976). In those interviews it was revealed that on several occasions questions by the King on specific details of the operations under review caught the leadership of the concerned ministry as well as the project manager by surprise.

21. Based on interviews with government officials (Kathmandu, 1976). The proposal was originally initiated by the Evaluation Section of the NPC and involved deliberations with the ministries and the palace. The NPC apparently did not succeed in mobilizing sufficient support for this proposal to allow the pilot-test to take place.

22. Based on interviews with government officials in several different ministries. It appears that in view of the already existing resentment toward the Ministry of Finance the proposal was shelved for the time being. In fact, all recognized that lack of cooperation on the part of the ministries would undermine the whole purpose of the proposed measure, which could affect other aspects of the relations between Finance and the other ministries.

23. See, J. C. Beyer, Budget Innovations in Developing Countries: The Experience of Nepal, Praeger Publishers, N.Y., 1973, pp. 16-

32. See also M. K. Shrestha, Trends in Public Administration in Nepal. Dept. of Information, HMG, Nepal. 1969. pp. 35-41 and 83-91. and, A. Wildavsky, "Why Planning Fails in Nepal." Administrative Science Quarterly, 17 (December 1972). pp. 516-520.

24. A. Wildavsky, Ibid., pp. 518-519.

25. In the road projects where such activities like filling road beds, compaction, early stages of culvert and bridge construction were not completed prior to the rainy season, it became necessary to do some of the work over again. Though the problem was not unique to the Nepali undertakings, the fact that they were more affected by the cumbersome accounting process caused a higher frequency of cases in which work could not be completed on time or according to plan, to avoid the rainy season.

26. CEDA, "Study of the Transport Corporation," op. cit. pp. 5-6.

27. Wildavsky identified 13 steps that were required to obtain quarterly release of funds to projects that involved a foreign aid component. See, "Why Planning Fails . . ." op. cit. pp. 519-520. It should however be noted that with the recent transfer of the programming section to the Ministry of Finance, some of the bottlenecks identified in that 13-steps process were apparently removed.

28. P. S. J. B. Rana, Nepal's Fourth Plan—A Critique. Yeti Pocket Books, Kathmandu 1971. pp. 31-34.

29. For a discussion on the administration of public enterprises like the NTL see, M. K. Shrestha, Trends in Public Administration in Nepal. op. cit. pp. 98-116.

30. On the basis of interviews held in Kathmandu with various government officials it became apparent that corruption in procurement involved orders for the purchase of equipment that was not needed for the projects. Given the de facto lack of effective control, this could be done and the persons involved were remunerated by the supplier for placing such orders. Due to the nature of the matter, the accuracy of these allegations could not be verified. However, it appears as though most projects had equipment that could not be justified by need. Furthermore, unverifiable allegations were also made that the NTL employed similar practices.

31. Based on interviews with officials who were familiar with the matter (Kathmandu, 1976).

32. CEDA, "A Study of the Transport Corporation," op. cit. p. 9.

33. To cite only a few examples see: A) L. J. Kroeger, "A Public Administration Program for His Majesty's Government of Nepal" Griffenhagen-Kroeger, Inc. Consultants. San Francisco. March 1962 (A report to USAID and HMG, Nepal). B) R. L. Podol, "An Approach to Improving Public Administration in Developing Countries," A report to USAID, Nepal. Kathmandu, March 1967. C) HMG, "Internal Administration and General Office Practices," A survey report prepared by the Administrative Reform Division. HMG. Kathmandu, February 1968. D) L. D. Bomberger, "The O & M Function in HMG" Past, Present, Future." Report to USAID, Nepal. Kathmandu, September 1974.

34. Based on interviews with government officials familiar with the subject matter. In some of those interviews it was mentioned that the practice of the newly appointed engineer forced even the American advisor to protest the lack of progress in the project, something that he would not consider doing under any other circumstances. Kathmandu, 1976.

35. "Engineering Evaluation . . ." USAID, 1973. op. cit. p. 48.

36. CEDA, A Study of the Transport Corporation" op. cit. p. 1.

37. Ibid., p. 8.

38. Ibid., p. 2.

39. Ibid., p. 4.

40. J. S. Gallagher, "Seventh Report to HMG Nepal and the UN Office of Technical Cooperation." Kathmandu, July 1972 (covering the period January-June 1972) p. 6.

41. Based on interviews with different government officials familiar with that subject matter (Kathmandu, 1976).

42. For in-depth discussion on personnel problems and organizational difficulties as well as a detailed proposal for changes required see, UNDP, "Road Feasibility Studies, Construction and Maintenance: Project Findings and Recommendations," N.Y., 1975 (DP/UN/NEP-69-516/1). pp. 31-67 and Annexes 5, 8-14 (pp. 81-110, 120-226). For a discussion on the issue of training administrative and technical personnel see, M. K. Shrestha and G. R. Agrawal, "Needs and Priorities in Training and Research for Development in

Nepal." Kathmandu, June 1975 (a study commissioned by the UN Asian Institute for Economic Development and Planning, Bangkok, Thailand).

43. "Engineering Evaluation . . ." USAID, 1973. op. cit. pp. 35, 41-42.

44. UNDP, "Road Feasibility Studies . . ." N. Y. 1975. op. cit. pp. 31-67, 81-110, 120-226.

45. Confidential report, prepared in 1970.

46. Confidential report, prepared in 1975.

47. H. O. Rieger and B. Bhadra, "Comparative Evaluation . . ." op. cit. pp. 213-214 (British project), 221 (India project), 228 (D-D), 230 (Russian project).

48. In the beginning of the D-D project practically no workers came from the local districts. Until 1973 the majority of the laborers came from the eastern and mid-western regions of Nepal. Similarly, in the early stages of the Kathmandu-Pokhara road project many laborers were brought in from the Kodari area, whereas in the Russian project laborers were brought in from districts like Birganj and Hetaura. Ibid., pp. 217 (Chinese), 225 (D-D), 230 (Russian).

49. See in particular the British project, though it could be found in other projects as well; Ibid., p. 212.

50. Though no specific details could be gathered on these allegations, they are widely believed to be true. It involved alleged corruption of high level officials who, in cooperation with big contractors, shared the income generated by the awarding of contracts. Several of the people interviewed claimed that approximately 25% of the total cost of each contract went to private pockets, while many others suggested the percentage was closer to 15%. At any rate, efficient execution became secondary and sometimes marginal to the goal of obtaining control over resources. On several occasions this also led to default by contractors who simply fled the project site, leaving work uncompleted (D-D project). Kathmandu, 1976.

51. For the detailed proposal see, UNDP "Road Feasibility Studies . . ." N.Y. 1975 op. cit. Annex 12 (pp. 193-197).

Chapter 7

THE PROCESS OF IMPLEMENTATION AND THE EXTERNAL ARENA

Implementation and the Challenge of Foreign Aid Projects

Cooperation between the donor and recipient country is an essential requirement in development assistance projects. It has, however, proved extremely difficult for both parties to attain their mutual as well as separate objectives. In fact, it appears as though aid donors and recipients often find themselves separated by common projects. This phenomenon, although presumably more probable in cases where the parties differ in their respective objectives, is also found in those instances in which the participants share common objectives. Differences in modes of operation and administrative techniques, differences in the criteria used to evaluate performance, and differences in the values and norms that guide the actors, present the parties involved with considerable obstacles to overcome. Furthermore, although attaining satisfactory cooperation in a given project may be an end in itself, it is usually perceived as a precondition to accomplishing the "altruistic" and "egoistic" objectives of the participants. Donors and recipients alike utilize development assistance projects as a means to attain various interests. Accordingly they seek to retain a measure of independence and flexibility. In this context, the requirement for cooperation may be seen by each party as a constraint upon its ability to pursue its own interests. The dilemma which thus confronts each participant is how to secure cooperation while retaining sufficient independence to pursue its own interests.

To the aid donor country executing a development assistance project, this dilemma poses a three-fold challenge. First, there must be organizational flexibility and adaptability to local conditions. Differences between the donor's environment and that of the recipient usually require the donor's executing agency to adjust its standard operating procedures to prevailing local conditions. Second, to establish a balance within the project's environment, the donor must find the optimal way to mobilize the logistic support needed without being drawn, albeit unwillingly, into local feuds. Third, a balanced relationship with the recipient requires that the donor seek to maintain sufficient independence, to accomplish its own interests, and to avoid antagonizing its partner-client, the recipient. The relationship between a donor and a recipient is based on the benefits expected to accrue to both parties from their cooperation. In this context there is rarely a situation of one-sided dependence where the recipient is dependent upon the donor, but not vice versa.

In Nepal, five aid donor countries carried out all the major road projects. Each donor sought to accomplish its own interests by its participation; each employed its own administrative methods and construction techniques; and each responded in a different way to the dynamics of the administrative process in Nepal. Consequently, five different patterns of road project execution evolved in Nepal and accounted for the manner in which the country's road network was established. Moreover, these patterns represented five different ways to deal with the aforementioned challenge as well as five different modes of interaction between the internal arena and the external arena. The discussion below describes these patterns and examines their implications on project execution.

Road Project Management—Five Profiles

The Chinese Pattern

China approached the construction operations with the premise that the recipient should be given the status of an equal partner. At the same time China sought to maintain control over all aspects of implementation and was particularly careful not to place Nepali officials in a position where they could or would be required to make decisions pertaining to execution. While for the first Chinese road project in Nepal--the Kathmandu-Kodari—some Nepali engineers of higher grades were employed, in subsequent projects—the Naubise-Pokhara road and the Kathmandu Ring Road—only junior engineers were assigned.[1] The Nepalis apparently understood that their status of equal partner did not imply an equal share of power and responsibility. Nepali personnel were employed mainly in administrative capacities while technical control and supervision were entirely in the hands of the Chinese. The Chinese projects were managed by a Team Leader with a Chief engineer and two Deputy Team Leaders under him. Additionally, each project was divided into several subdivisions, headed by a Chinese group leader with a staff of 15-20 Chinese technicians (engineers and overseers) under him. The Nepali personnel consisted of a chief engineer and an administrative manager, while in the subdivision there was one Nepali engineer with a few overseers, accountants and administrators under him.[2] On the Chinese side, however, the Team, Deputy Team and Group Leaders were not necessarily engineers.

Laborers on the Chinese projects were recruited locally, in the villages and through "gang" leaders. All were Nepalis. The Chinese, who considered the contract method incongruent with their ideology, introduced payment incentives for individual workers in the form of a "piece-work" system, and employed a similar approach in those few contracts that were awarded.[3] Furthermore, by bringing in a large number of middle level staff (i.e. Chinese engineers and overseers in the various subdivisions), the Chinese were able to maintain close supervision and teach the workers efficient work methods. As a result, performance improved and timetables were met more satisfactorily. An element of particular significance was the fact that most of the Chinese middle level staff spoke Nepali and maintained a life style quite similar to those of their workers.[4] Additionally, unlike most other donors, the Chinese imposed fewer demands on the gang leaders (who usually recruited laborers), thus enabling more Nepalis to take part in the operation. Most of the equipment and material were brought in from China, and the donor had its own service section responsible for equipment maintenance. In brief, through a combination of close supervision by skilled middle level personnel, the elimination of dependence on the Nepali administrative machinery, and the utilization of adequate methods for labor recruitment and payment to workers, the Chinese overcame some of the major obstacles in Nepali road construction operations.

The Indian Pattern

India has been the most active aid-donor in Nepal, and in the transport sector alone is responsible for the construction of five major roads. Project organization was based on the prevailing pattern in India. The personnel in charge of these undertakings were from the Indian Public Works Department (PWD). Only a few Nepalis participated in construction.[5] Each project was headed by a Superintendent Engineer who usually had between 4 to 10 Executive Engineers (one or two of whom were Nepalis from HMG's Roads Department) under him. The Executive Engineers headed the various divisions into which each project was divided, and each of them had between 10 to 15 junior engineers working under him. Thus, Nepali participation in the Indian projects was limited, although the Nepali Executive Engineers who were involved in the operation had similar responsibilities to those of their Indian counterparts.[6]

Unlike the Chinese pattern, most of the construction work in the Indian projects was awarded on contracts (of up to Rs 3 Lakhs at a time). Additional contracts were awarded to contractors who satisfactorily completed at least half of the first contract. When the department was unable to contract out segments of the construction on satisfactory terms, it carried out the work itself.[7] The supervision of the contractors and quality control of output was undertaken by the junior engineers of each division.

Table 4

A COMPARISON OF THE DONORS' PATTERNS OF ORGANIZATION

	Formal Status Accorded to Recipient	Staff in High Level DM* Positions	Supervisory & Mid Level Positions	Responsibility for Financial Matters	Responsibility for Equipment & Maintenance	Construction Work
China	Equal Partner	Chinese	Chinese	Chinese	Chinese	Chinese, Local Labor, Gang Leaders
India	Customer-Recipient	Mostly Indian	Mostly Indian	Indian	Indian	PWD & Private Contractors (Mostly Indian)
Britain	Customer-Recipient	British	British	British	British	Contractors, Gang Leaders
USSR	Partner-Contractor	Mostly Russian	Russians & Nepalis	Mostly Russian	Mostly Russian	HMG, Private Contractors
USA	Leading Partner	Nepali	Nepali	Nepali	Nepali	HMG, Private Contractors

*Decision-making

The contractors, most of whom were from India, were paid a lump sum for their work and were responsible for recruitment, organization and supervision of labor. Most contractors brought in Indian workers, particularly skilled laborers.[8] Additionally, most contractors had the necessary equipment at their disposal or rented equipment from the PWD (on an hourly or daily basis). Nepali laborers were of course employed by Indian contractors, but the latter seemed to prefer Indian workers whenever possible.

The projects' finances were controlled by Indians and the Nepali role was limited to HMG's Ministry of Transport approval of the estimated cost of the project. Following this approval, the amount was deposited in an Indian bank under the project name, and only the Executive Engineers could draw from that account.* The accounting system employed was the one that prevailed in India. The Indian pattern differed from the Chinese pattern in almost every respect, except in the extent to which both secured independence from the Nepali administrative machinery.

The British Pattern

In the British project, all key personnel were from the donor country. The Department's Chief Engineer was in charge of the entire undertaking and a Resident Chief Engineer was responsible for daily operations.** Under the latter were three main sections: A senior Quantity Surveyor who headed the work measurement department which dealt with the preparation of tender documents, contract awards and quantity measurement. Three Senior Engineers heading various divisions of the technical section (design, construction and supervision) and a Senior Executive Officer in charge of the administrative section (including accounts).[9] There were no Nepali officials in positions of responsibility. The Nepali engineers were new to the profession and participated as trainees.[10]

The labor force was recruited and organized in various districts adjacent to the project site, as well as from India. Two methods of labor recruitment were employed in Nepal: Recruitment through private contractors and gang leaders, which were most common; or through the Road Department who, in coordination with the Home

---

*In practice, some portion of that amount was deposited in a Nepali bank for the payment of local currency components.

**The Department's Chief Engineer was based in London and inspected the project twice or three times a year.

Ministry and the local Panchayats, recruited labor from villages along the alignment. It was, however, only in 1973 that most of the construction work was given out on contract.[11] Under the contract system, laborers were paid on a monthly basis and the gang-leaders were responsible for recruiting the work force. Local workers were hired on a daily basis, mostly unskilled laborers who preferred to work for short periods (2-3 weeks) and return to their villages. Equipment maintenance as well as the responsibility for project finances were in British hands and accountability was to the British authorities. Thus, the British, too, attained substantial independence from the Nepali administrative machinery.

The Russian Pattern

Russia constructed one major road in Nepal--the Dhalkewar-Pathlaiya section of the East-West Highway. The Russian staff was headed by a Chief Engineer and several senior Russian officials were in charge of the various sections of the project—a mechanical section (service and maintenance of equipment), an engineering section (design and detailed specifications) and a civil section (control and supervision of actual construction). The Russians hired the services of HMG's Roads Department as a general contractor responsible for the actual construction work. A second class engineer was placed in charge of the Nepali staff (engineering section, accounts and administration sections).[12] Still, the donor country reserved the right to award individual contracts to private contractors without consulting HMG.[13]

In addition to coordinating the activities of the Russian project-management with the departments of HMG, the Roads Department was responsible for the recruitment of labor. Most of the workers were brought in from neighboring districts, although groups of Indian workers were contracted specifically for certain types of work. Later, a system of petty contracts was also implemented. Given the mechanized techniques employed by the Russians, the overall number of laborers working on the project was relatively small.[14]

Equipment maintenance was the Russians' responsibility, while financial matters were handled jointly by the Russians and the Roads Department through a series of contracts. In effect, the donor and the recipient entered into separate contracts for the execution of different parts of the operation, and the Roads Department was paid according to the volume of work it completed on various sections of the road. Thus, with the exception of the Americans, the Russians depended most heavily on the Nepali administrative machinery in the execution of their road project.

The American-Nepali Pattern

The US pattern was designed to give HMG a central role in all stages of the operation. The US and Nepal agreed upon a division of labor in which HMG's Roads Department would be responsible for the entire D-D road project whereas USAID's involvement would include the financing of the project (75%) and the services of a technical advisor.[15] Nepali officials, headed by a senior engineer, managed the main sections of the project (planning, design, administration, accounts, technical-mechanical) and its subdivisions. The construction work was given out on contracts and the contractors themselves were responsible for labor recruitment. The workers for the D-D road project were recruited throughout Nepal as well as from India. The management was given the authority to deal with financial matters but a ceiling was placed on the amount of money it could approve. USAID renewed its contribution annually and could, therefore, exercise some control over expenditures, although it usually preferred to maintain a low profile. In any event, the USAID reserved the right to conduct independent evaluations on the progress of the project. In sum, the American-Nepali project was a joint operation wherein the recipient exercised considerable discretion.

Implementing Road Projects—The Donors' Record

China

Among the donors, China alone completed its projects within the original timetable. The Chinese first conducted a multi-faceted examination of prevailing conditions in Nepal and their projected impact on operations. China preferred to adhere to its own Standard Operating Procedures (SOPs) and practices; however, an attempt was made to adopt and apply Chinese SOPs to conditions in Nepal. In fact, China combined an emphasis on cooperation with considerable independence in operations and functional adaptability to prevailing conditions in Nepal. While following its own SOPs, China managed to make this practice acceptable to the recipient. Through careful and detailed preparation, China could confront the various problems which occurred during execution, providing ready-made and often suitable solution. This enabled the donor to impress the recipient with its efficiency, dependability and professionalism.[16] Nepal's recognition of China's willingness and capability to solve these problems enabled the latter to retain its control over the operation without seriously offending the idea of cooperation. Indeed, this approach was applied to most aspects of the operation—supervision, ordering, supply, recruitment and training of workers, maintenance and organization.

In their operations, the Chinese also employed "error-detecting devices" to correct for mistakes in the planning process and to cope with unexpected difficulties. This they accomplished by placing Chinese officials in various positions of the project's hierarchy. Consequently, China was able to secure both the capability to exert close control over operations and the capacity for early detection of errors. Moreover, these built-in and well-placed "devices" provided the project management with its own independent sources of information. China, therefore, was less dependent on either the Nepali system or private contractors for obtaining the necessary data on progres and activity during the course of the project. Information was also received from both Nepali officials and the independent contractors, all of which further solidified an effective system of control. Indeed, China overcame most of the difficulties encountered in the execution of road projects in Nepal, by devising a pattern of operation that emphasized adaptability to local conditions and the projects' environment as well as operational autonomy.

India

Due to its geoeconomic position in relation to Nepal, India enjoyed a considerable advantage over all other donors. Since India controls the gates to Nepal and since Nepal is dependent on the Indian market, India could easily adopt the pattern of cooperation that suited it best. Furthermore, compared to all other donors, India could absorb Nepali criticism without making drastic changes in its mode of operation. In executing road projects, however, India often failed to exhibit efficiency but succeeded in attaining independence and operational control, while limiting the involvement of the Nepali administrative machinery. The planning of Indian projects proved deficient, not only in comparison to China's performance but also in comparison to similar projects carried out in India. Detailed programs were deficient and several obvious constraints and difficulties remained unresolved. The complex and lengthy administrative and accounting procedures which prevailed in India spilled over into its projects in Nepal and accounted for both inadequate control and inefficient execution. In choosing to operate through a large number of private contractors, the donor did establish a group of loyal workers but often found it difficult to accomplish essential work in an efficient manner. The collection of data from the contractors was slow and incomplete. Consequently, the management, lacking effective error-detecting devices, were usually slow to correct mistakes and resolve difficulties during the course of the operation. This, together with a limited capacity to achieve proper coordination between the various participants involved in the projects, resulted in deficient implementation. Moreover, since India was involved in several projects in Nepal, it occasionally changed the order of priorities assigned to the different undertakings. These changes, which emanated from fluctuating conditions in India or in its relations with Nepal, led to additional delays. These delays in turn often undermined execution and resulted in low quality roads requiring frequent repairs.

Britain

Britain considered independence from the Nepali administrative machinery a prerequisite to professional performance but failed to consider those foreseeable difficulties which might seriously affect operations. This was particularly evident in the manner in which the project management dealt with local labor. The lack of satisfactory working conditions, compounded by insufficient consideration of local sensitivities in the hiring of Indian contractors and laborers, led to labor unrest and, therefore, occasional delays. Furthermore, since Britain worked through numerous private contractors, a great deal of coordination was required to attain satisfactory progress. This proved time consuming and perpetuated further delays and setbacks. Britain, however, enjoyed a certain advantage over several other donors which enabled it to sustain its independence from the Nepali system and to overcome various obstacles which impeded implementation. Due to a well-established network of contacts in the Indian market, Britain had a hinterland to fall back on for the provision of raw material and equipment. This advantage proved to be of considerable importance during 1974 when shortages of cement and oil brought many projects in Nepal to a halt. When the Nepali agencies were unable to obtain this materials, British officials used their contacts in India to secure supplies.[17] Without this advantage, Britain would conceivably not have been able to complete the project in 1976 (only two years beyond the original target date).

USSR

The USSR attempted to combine control over operations and an emphasis on cooperation with built-in mechanisms to neutralize operational constraints. It designated the Roads Department as its general contractor, assuming that a local agency would be better equipped to deal with issues like labor recruitment, coordination between Nepali organizations and the supply of construction material. In practice, however, the Russians found themselves more dependent on the Roads Department than they expected. Although Soviet personnel controlled most aspects of the operation, they could not operate without the assistance of the Roads Department, which was responsible for the actual construction work. The supply of spare parts for Soviet-made equipment proved a complex task, since these could not be easily obtained from India and the Russians failed to maintain an adequate stock on site. The Russians exhibited less ingenuity than the British in resolving such problems and demonstrated only limited flexibility and adaptability to local conditions.[18]

In 1971, when project costs rose considerably above estimates and delays had gotten out of hand, several changes were instituted. The USSR reduced the role of the Nepali Roads Department and began awarding work to private contractors. From 1971 to 1973, 45 kms of black topped road (out of the total of 110 km), 13 major bridges (out of 23), 25 medium and small bridges (out of 68) and 34 culverts and 14 causeways (out of 74 and 23 respectively) were completed.[19] This increased efficiency on the part of the Russians was due to both the changes they introduced in the project organization and operating procedures and to the distinct emphasis they placed on the necessity of completing the project. After five years, the Russians had finally accumulated a sufficient stock of spare parts for their equipment and they had gained the necessary experience to conduct a more successful project in Nepal.

USA

In view of the emphasis placed by the USA on recipient autonomy , issues such as the donor's adaptability to local conditions presumably became irrelevant. However, given the limited progress achieved in the D-D project, the suitability of this approach is highly questionable. The USA not only failed to take the necessary steps to ensure the success of the cooperative venture, but also demonstrated limited adaptability and organizational flexibility in coping with various aspects of the administrative process in Nepal and the interaction between HMG, the USAID mission in Kathmandu, and AID headquarters in Washington. The USAID mission in Kathmandu was obliged to abide by AID guidelines and procedures and HMG was forced to meet certain bureaucratic requirements which were actually beyond its capabilities. During the annual negotiations for the renewal of funds, for example, USAID's standard operating procedures required HMG to provide a variety of documents containing detailed information on various technical, administrative and economic aspects of the operation. In most cases, HMG lacked both the data and the capacity to collect and process this information, and was, therefore, obliged to admit its own inadequacies, to provide only partial information, or to resort to the fabrication of data. The donor, although it recognized the shortcomings of the documents submitted, failed to change its requirements. This created a source of embarrassment to the recipient which often led to an atmosphere of uneasiness and tension in HMG's dealing with USAID.[20] By failing to consider the broad range of implications inherent in its approach, US/Nepal relations were adversely affected. It should, however, be noted that the organizational arrangement applied in the D-D project was not repeated by the US in other projects in Nepal.

Implementation and Interaction between Donor and Recipient

In most cases, the modes of operation adopted by the donors did not resolve the difficulties inherent in development projects in Nepal and failed to strengthen the recipient's performance and capabilities. Due to limited organizational flexibility (with the possible exception of the Chinese) and limited Nepali participation in development projects, the donors increased the complexity of the implementation process and its chances for failure. Nepali officials were denied a fair opportunity to develop a proper understanding of operations and, therefore, could not contribute to improved performance. The dynamics of the internal arena were influenced by the availability of external aid in an environment that suffered from acute shortage of resources. The sense of commitment on the part of the Nepali agencies was reduced by improper handling of cooperative ventures (USSR, US) and by the exclusion of Nepalis from the planning and implementation of projects (India, Britain, China). The donors had their own interests to protect and, therefore, had to consider the possible effects of Nepali participation on smooth execution. However, neither the donors nor the recipient devoted sufficient consideration to the ways and means in which the execution of the project could contribute to the improvement of Nepal's development capabilities. Hence, the interaction that eventually evolved between the "internal arena" and the "external arena" could be characterized as unsatisfactory—both in terms of actual progress in construction operations and insofar as the development of indigenous capabilities.

Coordination and Communication

The difficulties involved in coordinating project participants were caused by both the donors and HMG. Nepal insisted on bilateral agreements with each donor as a basis for foreign aid activity in the country, and sought to ensure that on matters pertaining to the execution of projects, each donor would maintain contacts with the recipient agencies. Yet, it failed to consider the matter in terms of the administrative operations and requirements involved in such a commitment. Nepal, on the one hand, did not have the capability to provide the donors with the administrative and logistical support they required for efficient execution of projects. The donors, on the other hand, did not make the task any easier for the recipient since each preferred to employ its own SOPs in the execution of its projects. Each donor, with the possible exception of the US, sought to maintain operational control and expected the recipient's administrative machinery to assist in carrying out the project, but in a manner the donor deemed appropriate. Consequently, a situation evolved in which Nepal found itself obliged to deal with a variety of administrative systems, the requirements of which it could not possibly satisfy. The donors found themselves denied satisfactory assistance from the Nepali administration and were unable to attain sufficient independence from it. Furthermore, some Nepali officials and agencies realized that the power to delay projects was practically the only available means they had to further their particular interests. Hence, the timely provision of supplies, the issuing of equipment shipment documentation and the organization of an adequate labor force were not only factors which might create bottlenecks in coordination, they were also tactics employed by various agencies for leverage and control. While the Nepalis sought to exploit situations over which they had discretionary power, by so doing they forced the donors to seek solutions which side stepped the local bureaucracy. Consequently, dissatisfaction became a common feature in all projects, hampering coordination, communication and implementation at all levels and stages.

The political component inherent in foreign aid programs and the rules which governed the Nepali bureaucracy undermined communication between the donors and the recipient. Some donors exhibited considerable caution in their communications with Nepali agencies, preferring to avoid the issue of project difficulties which might be interpreted as undue or unacceptable criticism. Other donors did not consider it their duty to maintain frequent communication with the recipient, often choosing to provide reports only when necessary. Those donors who sought to maintain substantive communication with the recipient soon gave up as they realized that the recipient was either unwilling to reciprocate or unable to process the data and respond to it. Each donor eventually concluded that the most effective course of action was either to act without waiting for the recipient to make a decision, or to bring the matters they considered most urgent directly to highest administrative authority (i.e. the minister in charge or the secretary). Consequently, routine communication between the donors and the recipient was largely nonexistent.

The recipient, on the other hand, had only a limited capacity for absorbing large volumes of information, processing and properly disseminating it. Furthermore, the Nepali were often unwilling to maintain substantial exchange of information with the donors and to respond to their communications. They frequently avoided serious examination of reports from donors, let alone any follow-up procedures. Those donors who depended on the administrative support of the recipient—i.e. the USA and the USSR—found this tendency on the part of the recipient frustrating and counter-productive.

Indeed, it was against this background that the USAID commissioned the American team of consultants to conduct a study on the D-D project. The terms of the Task Order issued by USAID to this team clearly indicated that the donor sought to overcome the lack of data on the status of the project and on the prospects of completing it.[21] In the Soviet project, it was apparent that little data processing was performed during construction, at least as far as the Nepali agencies involved in the project were concerned. A study that dealt with this issue indicated that although some data was available in the Nepali project office, it was neither sufficiently complete nor systematic enough to provide an adequate picture of the operation. Moreover, this study noted that the bulk of the project records had been shipped to Moscow once the project office was closed.[22]

In projects where the donors were less dependent on the Nepali bureaucracy, the situation was somewhat different. In the Chinese projects, the data on various construction activities were rather detailed (with the exception of equipment input) and were made available to the Nepalis. In fact, the Chinese apparently maintained open channels of communication with the Nepalis.[23] However, in view of the fact that they controlled practically every aspect of the operation, communication between the donor and the recipient was largely a discretionary gesture. The recipient received ex post facto reports on those items considered appropriate by the donor. Furthermore, while China encouraged Nepali officials to maintain open channels of communication, in practice China responded only as it saw fit.

In the British project, the donor retained full control over on-site data, usually providing the Nepali authorities with brief progress reports, primarily in the form of annual reports. On those occasions when serious difficulties were encountered in the project, officials from the project and from the British embassy sought to communicate with the top echelons of the Roads Department or the Ministries of Transportation and Finance.

In the Indian projects, the availability of data pertaining to construction proved a problem to all concerned. Due to the fact that in the main the contractors themselves provided both the labor and the equipment but were less than diligent in maintaining records, the Indian officials lacked basic information on the progress of the operation. The Indian Cooperation Mission (ICM) in Kathmandu maintained more or less regular communication with the Roads Department but rarely did it provide HMG with comprehensive briefings on the entire operation or with sufficient data that would enable Nepali agencies to reach a comprehensive understanding of the project.

The implications of these patterns of communication on the execution of projects were substantial in several different ways. Inadequate and insufficient communication between the participants reduced the prospects of correcting operational deficiencies. It prevented the establishment of a system that could provide both sides with potentially effective mechanisms for early detection of problems. Hence, operational difficulties were usually dealt with only when they became serious. Second, HMG failed to establish a system of reporting that would allow it to track activities in various projects effectively. Moreover, it never developed a proper understanding of the different operations that were undertaken in various parts of the country. Thus, due to ineffective communication, both the donors and the recipient committed various mistakes which further complicated operations and increased the dissatisfaction of both parties.

In that respect, the recipient was also denied a fair opportunity to apply the lessons learned in the various road projects to other undertakings in other fields and other parts of the country. HMG therefore repeated the same mistakes in different projects and remained unable to improve its own operations and provide prospective donors with sufficient background information on conditions pertaining to ongoing or proposed projects. The study conducted by CEDA on "Comparative Evaluation of Road Construction Techniques in Nepal"[24] (1973) was an outstanding exception to the general state of affairs and an example of what could and had to be done in the realm of communication between the donors and the recipient to encourage more efficient execution of projects.[25]

The "Indian Factor"

To paraphrase a Kathmandu cliche—"When India catches cold Nepal sneezes." Thus, Nepal's development policies cannot be evaluated only on a donor-recipient basis, but rather, with due consideration to the "Indian Factor." In the transport sector and in road projects in particular, Nepal's effort to establish its own transportation network was greatly influenced by its geoeconomic position with respect to India. India has on several occasions imposed restrictions on the export of commodities required for projects in progress as a means of expressing its dissatisfaction with Nepal's policies.[26] In 1971, for instance, during the months that preceded the signing of a new Trade and Transit Treaty between the two countries, India imposed such restrictions. Consequently, serious slowdowns were experienced on all road projects, including some of India's own projects in Nepal. At various points in time India ranked some of its own projects as low priorities, or it simply did not consider delays in operations to be a serious deficiency.* In 1973-1974 the execution of road projects was undermined by another aspect of Nepal's dependence on India. The 1973 Arab oil embargo depleted India's reserves. Since India was the major supplier of petrol to Nepal, India's shortage was immediately felt on Nepal's road projects. Moreover, Nepal found it highly complicated and costly to obtain oil from other sources and therefore had to accept a slowdown in operations—which led to increased costs in most projects.

An additional component of the "Indian factor" was Nepal's dependence on India's transport system. Port storage and facilities available to Nepal in India were poor.[27] Improper handling of shipments and thefts were daily occurrences. Clearance limitations and complex procedures for the licensing and issuance of transit permits were immensely time consuming. Overcoming the difficulties of the overloaded Indian transport system was impossible, making delayed deliveries an inescapable fact. Consequently, the planning of specific work programs according to projected availability of materials and spare parts was usually an exercise in futility. These delays were experienced by all donors.

The unusually strong influence of private Indian companies over Nepal's transport industry also had direct bearing on transport services. Private Indian trucking companies held a virtual monopoly over the trucking industry in Nepal and determined freight costs. To maximize their own profits, these companies concentrated on servicing a few important markets close to the Indian border.* These companies refrained from establishing service to other parts of the country and allegedly insured their monopoly by adopting rather questionable practices. The lack of adequate railway services and the unsatisfactory operation of the ropeway in Nepal were party attributable to the pressures and incentives brought to bear by Indian companies.[28] Furthermore, due to the fact that these private

---

*In fact, India did not even try to exploit the shortages in other donors' projects to its advantage.

companies provided Nepal with an essential services, the country was placed in a delicate situation wherein it had to accommodate the interests of these companies despite the fact that services were often rendered at inflated costs. In sum, the "Indian factor" affected various aspects of construction operations and was often the cause of additional difficulties in the execution of road projects.

Conclusions

The implementation of road projects in Nepal was a multifaceted process influenced by a broad variety of factors in the internal and external areas as well as by the complexity of interactions between them. The political and economic interests of the different participants greatly affected the country's development efforts, sometimes out of all proportion. The dynamics of Nepal's politicized bureaucracy and the different modes of operation employed by the foreign participants determined the manner in which operational difficulties were resolved; they also created problems that plagued the operation, compounded by conditions of poverty in the country. The execution of projects necessitated organizational flexibility which was dealt with in different ways by different participants. A variety of patterns of communications evolved among and between the participants, exacerbated by bottlenecks in the exchange of information. Various coordination efforts, different forms of administrative control mechanisms and diverse styles for management of manpower resources evolved in a similar fashion and affected the progress achieved on projects. Technical operations pertaining to equipment and maintenance activities constituted yet another facet of the interaction between the participants challenging tasks magnified by the geographic setting and climatic conditions in Nepal. The aid-donor patterns which evolved therefore represented not only the variety of possible solutions to problems in road construction undertakings but the unique demands presented by circumstances in Nepal, including the "Indian factor." This complex web of interactions determined the successes and failures of Nepal's national road network. In recent years, Nepal succeeded in resolving some of these difficulties and became somewhat more effective in executing its development programs. However, it is yet to achieve a breakthrough in the vicious circle of poverty.

---

*The Tribhuvan Highway, Biratnagar, Janakpur, Bhairawa, Sidhartha Highway, Nepalgung and Mahendranagar were the main areas serviced.

REFERENCES

1. H. O. Rieger and B. Bhadra, "Comparative Evaluation of Road Construction Techniques in Nepal" CEDA, Kathmandu. 1973. p. 215.

2. Ibid., Ibid.

3. The "piece-work" system was based on compensation for work completed. The worker received full payment even if he completed the work ahead of the allotted time. Contract payments were made in the presence of the laborers, thus reducing the prospects of workers being cheated by gang leaders. Ibid., pp. 216-217.

4. Ibid., p. 217.

5. The exception was the Tribhuvan highway which was built by the Indian Army Corps of Engineers.

6. Ibid., p. 219.

7. Ibid., p. 220.

8. Ibid., p. 221.

9. Ibid., p. 211.

10. Initially, senior engineers were assigned to the project, but after it became apparent that they were not about to be given any major responsibilities, they were replaced by junior engineers.

11. Ibid., p. 212-213.

12. Ibid., p. 228-229.

13. In the later stages of the project the Russians did indeed award parts of the work to private contractors in order to accelerate the speed of construction work.

14. Ibid., p. 230.

15. In 1973, following a study by an American team of consultants, the USAID also provided an additional advisor on equipment operation and maintenance.

16. Based on interviews held with several Nepalese officials who were involved in these projects. (Kathmandu, 1976).

17. Based on interviews held with British officials who were

associated with this project. (Kathmandu, 1976).

18. The Soviet personnel was allegedly inclined to avoid a speedy completion of the project and refrained from seeking immediate solutions to various difficulties that slowed construction work. These observations were made by several Nepali and foreign officials in interviews held in Kathmandu, 1976.

19. Rieger and Bhadra, op. cit. p. 103.

20. Both Nepali and USAID officials have mentioned this as an "American-Nepali" dilemma. Moreover, Nepali administrators have criticized the USAID for being insensitive to the matter. In this context, the recurring theme has been a comparison to China's ability to understand the issue and to act upon it properly (i.e. refraining from embarrassing the recipient).

21. "Engineering Evaluation . . ." USAID, 1973. op. cit. p. 2.

22. Rieger and Bhadra, op. cit. pp. 107-108.

23. Rieger and Bhadra. op. cit. p. 88.

24. A joint undertaking with ILO.

25. For a proposal on the kind of data needed and the actions that Nepal had to take in order to utilize more effectively the aid program available to the country see Rieger and Bhadra, op. cit., pp. 232-256. It should, however, be noted that this proposal dealt mainly with the issue of construction techniques without elaborating on the administrative arrangements pertaining to it or, the political dimensions of those issues.

26. For a discussion on the issue of Indo-Nepal Trade and Transit treaties and the political and administrative aspects involved see, Tribhuvan Nath, The Nepalese Dilemma, Sterling Pub. New Delhi, 1975 pp. 403-441 (this reflects primarily an Indian point of view). P.G. Lohani, "Nepal-India Trade Relations," CEDA Study Series (Occasional paper No. 1—Trade and Transit: Nepali problems with her Southern Neighbors) Kathmandu (no date). Mahammed Ayoob, "India and Nepal: Politics of Aid and Trade," The Institute for Defense Studies and Analyses Journal, Vol. 3, No. 2 (October 1970) pp. 127-156. See also Rishikesh Shah, Nepali Politics: Retrospect and Prospect, Oxford University Press London, 1975 pp. 132-137.

27. J. S. Gallagher, Jr., "Seventh Report . . ." op. cit. 1972. p. 9.

28. Rumors of widespread "pay offs" by the Indian trucking companies to Nepali officials in the NTC, the ropeway management and the ministry were a recurrent theme in almost all the interviews held in Nepal.

INDEX